Nuclear Geochemistry

Advances
in
Science
and
Technology

ADVANCES IN SCIENCE AND TECHNOLOGY

Nuclear Geochemistry

NATALYA A. TITAYEVA

Translated from the Russian by G. Egorov

Mir Publishers
Moscow

CRC Press
Boca Raton Ann Arbor Tokyo London

Library of Congress Cataloging-in-Publication Data

Titaeva, N. A. (Natal'ia Alekseevna)
[IAdernaia geokhimiia. English]
Nuclear geochemistry / N. A. Titayeva : translated from the Russian by G. Egorov.
p. cm. -- (Advances in science and technology in the USSR)
Includes bibliographical references and index.
ISBN 0-8493-7545-2
1. Isotope geology. I. Title. II. Series.
QE501.4.N9T5813 1994
551.9--dc20 94-6352
CIP

This edition published by CRC Press, Inc., 2000 Corporate Blvd. N.W., Boca Raton, Florida
Direct all inquiries to CRC Press, Inc., 2000 Corporate Blvd., N.W., Boca Raton, Florida 33431.

International Standard Book Number 0-8493-7545-2
Library of Congress Card Number 94-6352
Printed in the United States of America 1 2 3 4 5 6 7 8 9 0
Printed on acid-free paper

CONTENTS

INTRODUCTION

What Does Nuclear Geochemistry Study?

Nuclear geochemistry studies radionuclides and their related nuclear processes in the Earth. The Earth contains a host of radionuclides with a diverse genesis: primary natural radionuclides and their decay products, cosmogenic radionuclides, and radionuclides of anthropogenic origin. The specific properties of radionuclides make it possible for them to be used as tracers and geochronometers of various processes: from nucleosynthesis, the evolution of the crust-mantle system to current processes in the biosphere, atmosphere, and hydrosphere and on the surface of the lithosphere.

The properties of radionuclides of interest for the Earth sciences are:

1. Spontaneous decay at a constant rate characteristic of each group of radioactive atoms with the same value of A and Z. Atoms of this kind are called *nuclides*. Moreover, the amount of primary radioactive nuclide (or *radionuclide*), called the parent variety, steadily decreases. At the same time, in place of each decayed atom there appears a new one, which is called the *daughter nuclide*. This may be radioactive and disintegrate to a subsequent nuclide at a characteristic rate. Either a single act of decay, or a chain of disintegrating atoms, always ends in a stable nuclide. Thus, the decay of radioactive atoms is always accompanied by an accumulation of stable decay products. Since this process takes place at a constant rate, it becomes possible to determine the time that has passed since the parent nuclides entered the geological object by the decrease in the number of parent atoms or by the accumulation of daughter ones. This time is called the isotope or radiological age. Although sometimes it is impossible to obtain any genuine information on the age, the distribution of nuclides enables us to estimate the rate of a geological process.

Relatively short-term radioactive nuclides, the half-life of which is commensurable with current processes, are widely employed as natural tracers of 'young' processes. Nuclides whose life-spans exceed the age of the Earth itself may be tracers of either ancient processes or the slow global processes of the evolution of the Earth.

2. The use of radioactive nuclides as tracers of natural processes calls for knowledge of the distribution and behaviour of the nuclides in these processes. Hence, an important section of nuclear geology is the geochemistry of natural radionuclides. Most important is the geochemistry of uranium and thorium, which are the raw materials of nuclear energetics. No less important is the study of radionuclides of anthropogenic origin, which has so far not been adequately elaborated.

3. All radioactive transformations are accompanied by the emission of ionizing radiation. It is a well-known fact that radiation leads to changes in natural matter; in minerals, for instance, these changes can very often be accumulated during the time of radiation and can also be employed as geochronometers.

4. Radioactivity is the main diagnostic feature serving to discover radioactive nuclei. The measurement of radiation is used for the analysis of radionuclides in laboratory conditions. Field measurements are widely employed in the exploration of radioactive raw materials, quests of mineral products distinguished by their radioactivity from the enclosing rocks, the geological mapping of rocks with various content of radionuclides, etc.

5. Spontaneous radioactive decay is an exothermic reaction which is accompanied by the evolution of heat. The heat released in the process of radioactive transformations plays a substantial role in the Earth's energy balance.

The History of Nuclear Geochemistry

Research in the field of nuclear geochemistry began soon after radioactivity was discovered. Suffice it to say that Pierre and Marie Curie carried out their first experiments with uranium pitchblende ore. Many significant discoveries and ideas in nuclear geochemistry are connected with the name of Pierre Curie. In 1903 together with Laborde he discovered that the radioactive decay is accompanied by the release of heat proportional to the number of decayed atoms. 'This discovery was the first in geology to scientifically raise the question on the actual source of energy in geological processes' [1]. The geological significance of this discovery was duly appreciated by Strutt in 1906 and Jolly in 1908. The idea that it was possible to determine geological time by the radioactive decay of natural nuclei was also first suggested by Pierre Curie. Speaking at a session of French Physical Society in 1902, he noted that the constant rate of radioactive decay gives mankind the measure of time, its 'standard', independent of the processes taking place on Earth.

The practical possibility of determining the age of natural objects by the accumulation of decay products was noticed by B. Boltwood and E. Rutherford in 1904. Strutt carried out comprehensive studies to determine U, Th, Ra, and He contents in minerals and rocks. In 1908 Strutt showed that the He/U ratio depends on the geological age of minerals.

In Russia, Vernadsky was the first who comprehended the importance of investigating radioactive processes for geology. Vernadsky's concern with radioactivity manifested itself in 1908 after a report delivered by T. Jolly at the Congress of the British Association of Scientific Studies in Dublin on the role of radioactivity in the processes taking place on Earth. The report and the talks Vernadsky had with the British scientist produced a great impression on him and inspired him to organize studies in Russia on radioactivity in the Earth's depths. The goals of this new trend in geochemistry were set by Vernadsky as far back as 1910: 'What is actually presented has significance for the interpretation of the Earth's heat, the establishment of the age of our planet, the estimation of geological periods in years, and the formation of mountain chains; we can see its manifestation in the scientific study of atmospheric electricity, and the chemical reactions of the Earth's surface' [1].

A bit later he called the new branch of geology 'radiogeology', and gave it the following definition: 'Radiogeology studies radioactive processes on our planet, their influence, and manifestation in geological phenomena' [1]. Almost all the sections of radiogeology (nuclear geochemistry) and the research implemented at the initiative of Vernadsky retain their practical significance today.

The first task that he had set before his colleagues consisted in investigating radioactive minerals and their deposits in Russia. From 1908 Vernadsky had worked uninterruptedly to organize expeditions and a laboratory that would be concerned with explorations and the study of radioactive minerals. The first expeditions between 1908 and 1913 included K.A. Nenadkevich, A.Ye. Fersman, D.S. Belyankin, Ya.V. Samoilov, L.A. Kulik, V.I. Kryzhanovsky, B.N. Lindener, G.I. Kasperovich, E.D. Revutskaya, and others. The permanently functioning radium expedition of Academy of Sciences created in 1913 gave rise in 1922 to the State Radium Institute in Leningrad and its branch in Moscow. At that time it was the only organization dealing with research into natural radioactivity, where alongside the departments of physics and chemistry, there also existed the radiogeologial section. Vernadsky conducted the research on the radioactivity of minerals and rocks along two basic trends: (1) the distribution of radioactive elements in the Earth's crust with a view of compiling the radiogeological map of the Earth's surface, and (2) the quests of radioactive raw materials. The purpose of radiogeological map consisted in establishing the relation between geothermy and the structure of the lithosphere. 'It is quite clear that this kind of map should furnish the distribution of physically quite different and extremely important parts of the Earth's surface: the border regions, which are not identical in terms of temperature. Such border regions should be of significance for various geological and geophysical phenomena, and within these areas there should be new phenomena not found elsewhere' [1]. It is natural, therefore, that such a map can only be compiled after a vast amount of data has been accumulated on the concentration of uranium, thorium, and potassium in the rocks of the upper lithosphere, which is accessible for systematic testing. A radiological map of the USSR was only compiled at the end of the seventies at the All-Union Geological Institute in Leningrad by A.A. Smyslov, who generalized the research of many teams of workers.

Work on the radioactivity of rocks is connected with research into the thermal history of the Earth and current geothermy. The most prominent scientist in this field is E.A. Lyubimova. Works by A.A. Smyslov, B.Ya. Smirnov, B.I. Kononov, and B.G. Polyak are also noteworthy.

Prospecting in the first period was mainly to find radium, then the only source of radioactivity. Research carried out in the twenties and the thirties indicated the high content of radium in waters of oil beds enriched with calcium, sodium and chlorides (L.N. Bogoyavlensky, V.I. Vernadsky, V.I. Baranov, I.D. Kurbatov, V.V. Cherdyntsev, and others). Oil brines were processed industrially to extract radium [2].

The second stage in the development of nuclear geochemistry was connected with the quest for raw materials for the atomic industry, viz., deposits of uranium. During the thirties and the forties scientific studies concentrated on the granites which are richer in radioactive elements than other rocks (L.V. Komlev, V.I. Gerasimovsky, and others). Since the middle of the century the geochemistry of uranium has been closely studied, particularly the processes of ore formation. This research has been pioneered at the Vernadsky Institute of Geochemistry and Analytical Chemistry by A.P. Vinogradov, A.I. Tugarinov, G.B. Naumov, N.P. Ermolaev, V.L. Barsukov, and by scientists at other organizations (A.I. Perelman, A.K. Lisitsyn, V.M. Gavshin, B.I. Omelyanenko, A.A. Smyslov, E.V. Plyushchev, Ya.N. Belevtsev, R.P. Rafalsky).

The development of exploration geochemistry for uranium produced a new discipline which began to be applied in various fields of geology. This involved the ratios between the individual radionuclides in the natural decay series. This is the isotopic non-equilibrium method and it was developed theoretically by V.V. Cherdyntsev, who, in collaboration with his post-graduate P.I. Chalov, analytically proved the presence of isotope shifts in natural uranium for the ratio $^{234}U/^{238}U$. The first information on the effect was made public in 1953 at the third session of the Commission on the determination of the absolute age in Moscow [3]. The method was further elaborated by N.E. Syromyatnikov, K.E. Ivanov, V.I. Malyshev, N.A. Titayeva, and V.L. Zverev.

One of the most important trends in nuclear geochemistry is the development of radiological methods of geochronology, which in the country is closely associated with V.I. Vinogradov, A.I. Tugarinov, I.E. Starik, V.I. Baranov, E.K. Gerling, L.V. Komlev, N.P. Semenenko, and Yu.A. Shukolyukov. Contemporary nuclear geochemistry is characterized by a comprehensive development of isotope geochemistry, which makes use of isotope relationships as natural tracers of geological processes and the study of the geochemistry of radionuclides of anthropogenic origin and their use as radioactive geochemical tracer.

A specific trend in nuclear geochemistry which began to develop some time ago is the research into the radiation changes of the environment. Geochronological studies which make use of the dose-effect have been flourishing in Leningrad, at the Institute of the Geology of the Precambrian Period (Yu.A. Shukolyukov, A.N. Komarov), the Moscow State University (V.K. Vlasov, A.S. Marfunin, and A.I. Shlykov). The radiolytic nature of a large number of geochemical phenomena in underground waters were investigated by I.F. Vovk and L.N. Gutsalo.

REFERENCES

1. Vernadsky, V.I., 1954-1960. Collected Works. In five volumes. USSR Acad. Sc. Publ., Moscow.
2. Pogodin, S.A., Libman, E.P., 1977. How Was the Soviet Radium Recovered. Atomizdat, Moscow.
3. Cherdyntsev, V.V., 1969. Uranium-234. Atomizdat, Moscow.

1 Physical Properties of Radionuclides

The chapter is devoted to the basic concepts of nuclear physics pertaining to the nature of radioactivity and its properties. This information is essential for an understanding of later material, and is presented without any special mathematical apparatus.

1.1 General Information on Atomic Nucleus

The mass of atomic nucleus is over 99.95% that of the atomic mass. The size of nucleus is of the order of 10^{-12}–10^{-13} cm, whereas that of the electron shells is approximately 10^{-8} cm. Thus, the size of nucleus is at least 10^4 times smaller than the size of an atom. This shows that the radius of action of nuclear forces binding the particles within the nucleus does not exceed 2×10^{-12} cm.

It follows from the quantum theory that atomic nuclei have a level character of energy distribution. The existence of energy levels in the nucleus means that its energy spectrums are discrete. Atomic nuclei consist of elementary particles—protons and neutrons. The mass of a proton (M_p) and that of a neutron (M_n) are very much the same: $M_p = 1.67239 \times 10^{-24}$ g, $M_n = 1.67470 \times 10^{-24}$ g. It almost 2000 times greater than the mass of an electron (m_e): $M_p = 1836\ m_e$; $M_n = 1838\ m_e$. The proton is electrically positive. Its charge is equal to the charge of electron in absolute value. The electric charge of a neutron is zero.

The protons and neutrons entering into the composition of atomic nuclei are termed *nucleons*. The total number of nucleons in the nucleus is called the *mass number* and is denoted by *A*. The number of protons in the nucleus determines the nuclear charge in units of electron charge and is called the *atomic number* of a chemical element (*Z*). The number of neutrons in the nucleus is denoted by *N*.

Normally, the atom is electrically neutral, since the number of electrons in the atomic shells is equal to the number of protons in the nucleus. The chemical properties of an atom are determined by the structure of its electron shells and the number of electrons. Therefore, atoms with the same *Z* share the same chemical properties. Nuclei with the same atomic number *Z*, but with different mass numbers *A* are called *isotopes* of the chemical element. For instance, $^{1}_{1}H$, $^{2}_{1}H$, $^{3}_{1}H$ are isotopes of hydrogen.

Nuclei with the same *A*, but different *Z*, are called *isobars*. An example of a triad of isobars is $^{40}_{18}Ar$, $^{40}_{19}K$, and $^{40}_{20}Ca$.

Nuclides are groups of atoms with the same *A* and *Z*. Radioactive nuclides are also called radionuclides. The term isotope is used only together with the name of an element, e.g. 'Radionuclide $^{14}_{6}C$ is an isotope of carbon'.

The energy of the nuclear bond $\varepsilon_{Z,A}$ is the total energy required to split the atom into its components: protons and neutrons. It characterizes the stability of atomic nuclei and helps to estimate the energy balance for any nuclear transformations.

Instead of the binding energy, the value $\varepsilon_{Z,A}/A$ is often used. This is the specific binding energy, or the binding energy per nucleon.

We can apply Einstein relation to express the binding in terms of mass:

$$\varepsilon_{Z,A} = \left\{ M_{Z,A} - \left[ZM_{\mathrm{p}} + (A - Z)M_{\mathrm{n}} \right] \right\} c^2$$

where M_{p} is the mass of a proton, M_{n} is the mass of a neutron, $M_{Z,A}$ is the mass of the nucleus, and c is the speed of light. The nuclear mass is smaller than the sum of the masses of constituent nucleons by $\varepsilon_{Z,A}/c^2$.

For instance, an alpha-particle with $e_a = 28$ MeV is lighter than two isolated protons and two neutrons by $\frac{\varepsilon_\alpha}{c^2} \approx 4 \times 10^{-26}$ g, which makes 0.7% of the mass of the alpha-particles.

The binding energy expressed in atomic mass units is called the mass defect Δ (1 amu = 1/12 mass of $^{12}_{6}C = 1.6582 \times 10^{-24}$ g or 931.44 MeV).

The mass defect Δ is equal to the difference between the nuclear mass in the atomic mass units and its mass number:

$$\Delta = \frac{M_{Z,A}}{1/12 M_{^{12}\mathrm{C}}} - A$$

The specific binding energy is equal to about 8 MeV per nucleon. The relative independence of the specific binding energy of the number of nucleons is explained by the fact that each nucleon can only interact with neighbouring nucleons.

The maximum of specific binding energy (8.8 MeV) is situated in the periodic system near A = 56, i.e. in the region of iron. In terms of energy it is much more advantageous for light nuclei to fuse into heavier ones, whereas the division of a nucleus into lighter nuclei is more favourable for the heavier nuclei.

The mathematical surface $\varepsilon_{Z,N}$, as a function of Z and N, can split into three surfaces, the distance between which is about 1 MeV. The upper level is occupied by the surface where the even-even nuclei (even Z, even N) are situated. The median position corresponds to odd-even and even-odd nuclei (odd Z, even N, or the other way round). The lower surface is for odd-odd nuclei (odd Z, odd N). This illustrates the 'pairing' of nucleons in the nucleus. During pairing the binding energy in the nucleus increases by approximately 1 MeV.

Among the stable isotopes the even-even ones are preponderant, while the least number is with the odd-odd varieties, which are known to be 4 all in all: ${}^{2}_{1}H$, ${}^{6}_{3}Li$, ${}^{10}_{5}B$, and ${}^{14}_{7}N$. There exist particular 'magic' values of A and Z, for which $\varepsilon_{Z,N}$ gives rise to small clear-cut maximum. The magic values are equal to: 2, 8, 20, (28), 50, 82, 126 and correspond to the completed nuclear shells, i.e., to the structures similar to the electron shells of noble gases. Those sharing the maximum specific binding energy are the doubly magic nuclei: ${}^{4}_{2}He$, ${}^{16}_{8}O$, ${}^{40}_{20}Ca$, ${}^{208}_{82}Pb$.

1.2 Types of Radioactive Transformations

Radioactivity can be defined as the ability of atomic nuclei to decay spontaneously and emit one or several charged particles and photons. The nuclei that can spontaneously disintegrate are said to be radioactive, while those that cannot do so are called stable.

1.2.1 Beta Decay

Out of 1700 nuclides known to us at present only 200 are stable. Most of the radioactive nuclides have been obtained by artificial means. In the diagram of Fig. 1.1 in the coordinates (N, Z) the nuclides that are known to exist in nature are plotted. The stable nuclei designated by black circles occupy a narrow band. When the values of A are small ($\leq$ 40) the ratio N/Z = 1. When the total number of nucleons in the nucleus increases, the number of neutrons increases as well, because ever-increasing Coulomb repulsion of protons and the ratio reaches 1.6 for $^{238}_{92}U$.

The nuclei in diagram of Fig. 1.1 to the right and left of the stable nuclei band are said to be radioactive. The radionuclides of the left-hand part of the diagram are characterized by having more neutrons as regards the stable ratio

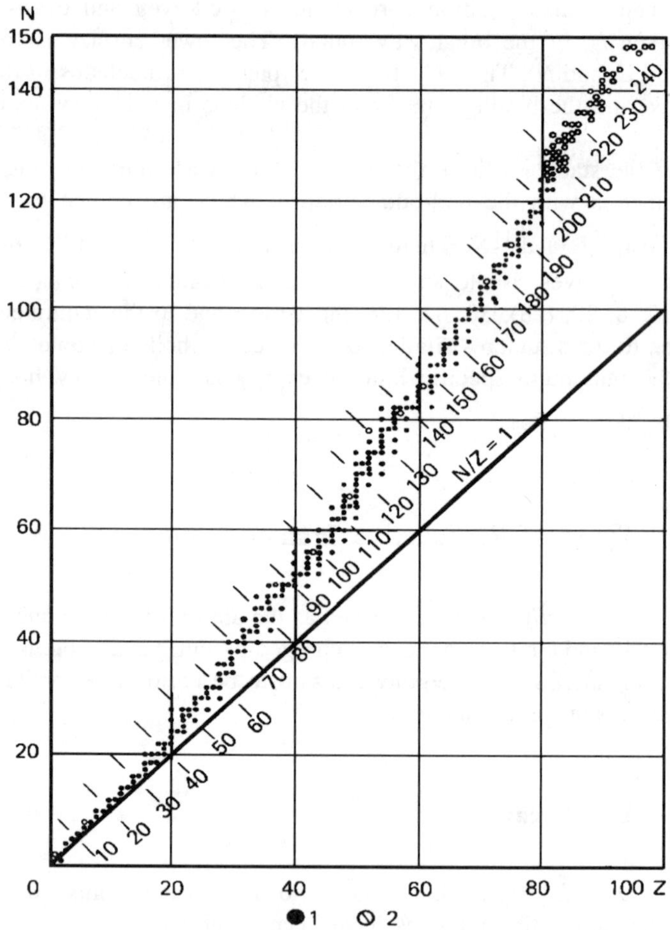

Fig. 1.1. Neutron-proton diagram:

1—stable nuclides; *2*—radioactive nuclides; *N*—number of neutrons; *Z*—number of protons

of $\frac{N}{Z}$ for the corresponding chemical element. They become transferred to the stable state by converting excess neutron into proton with the emission of electron (the beta particle β^-) and antineutrino:

$$ {}_Z^A X \rightarrow {}_{Z+1}^{A} X + \beta^- + \tilde{\nu} \tag{1.1}$$

$$(A_2 = A_1;\ Z_2 = Z_1 + 1)$$

This kind of nuclear transformation is called *beta decay*. Accordingly, a new nucleus is formed with the same mass number and the atomic number increases by unity. For instance,

$$ {}_6^{14}\mathrm{C} \rightarrow {}_7^{14} N + \beta^- + \tilde{\nu}$$

The radionuclides situated in the diagram to the right of the stable nuclei possess an excessive number of protons. For the transformation of the nucleus into a stable state one of the protons in this case must be converted into a neutron.

There exist two transformations of this kind.

(a) The nucleus emits a positron β^+ and neutrino:

$$ {}_Z^A X \rightarrow {}_{Z-1}^{A} X + \beta^+ + \overline{\nu} \tag{1.2}$$

$$(A_2 = A_1;\ Z_2 = Z_1 - 1)$$

This transformation is called positron or β^+-decay.

(b) The nucleus absorbs one of the electrons of atomic shell (usually from K-shell) emitting neutrino:

$$ \overline{e} + {}_Z^A X \rightarrow {}_{Z-1}^{A} X + \overline{\nu} \tag{1.3}$$

$$(A_2 = A_1;\ Z_2 = Z_1 - 1)$$

This type of transformation is called electron capture or K-capture.

In the case of positron decay, as in K-capture, the newly-formed nucleus does not differ from the preceding one as regards the value of A, though it has the atomic number one unit lower.

It is also interesting to notice how the stability of isotopes changes for the same element. As an example, it is possible to consider the isotopes of silicon with $Z = 14$:

$$\underset{\text{excess } Z}{^{27}\mathrm{Si}^*} \quad \underbrace{^{28}\mathrm{Si}\,^{29}\mathrm{Si}\,^{30}\mathrm{Si}}_{\text{stable}} \quad \underset{\text{excess } N}{^{31}\mathrm{Si}^*} \quad (^*\text{—radioactive isotope})$$

$$^{27}_{14}\mathrm{Si} \rightarrow {}^{27}_{13}\mathrm{Al} + \beta^+ + \overline{\nu} \quad \text{(positron decay)}$$

$$^{31}_{14}\mathrm{Si} \rightarrow {}^{31}_{15}\mathrm{P} + \beta^- + \tilde{\nu} \quad (\beta\text{-decay})$$

A necessary though not always a sufficient condition for the radioactivity of a nucleus is the energy profit of transformation: the mass of the parent nucleus should exceed the sum of the masses of the final nucleus and all the particles that are formed during disintegration.

In the equations of energy balance of β-decay, we do not take into account the rest mass of the neutrino and antineutrino, which is zero.

Hence, β-decay is energetically advantageous, if:

$$^{A}_{Z}M > {}_{Z+1}^{\;\;A}M + m_e \;\; (\beta^-\text{-decay}) \tag{1.4}$$

where m_e is the electron mass, $^{A}_{Z}M$ and $_{Z+1}^{\;\;A}M$ are the masses of the parent and final nuclei.

Positron decay and K-capture are possible in energy terms, if:

$$^{A}_{Z}M > {}_{Z-1}^{\;\;A}M + m_e \;\; (\beta^+\text{-decay}) \tag{1.5}$$

$$^{A}_{Z}M + m_e > {}_{Z-1}^{\;\;A}M \;\; \text{(K-capture)} \tag{1.6}$$

In the case of β⁺-decay and electron capture the same process of the transformation of proton into neutron takes place, i.e. these transformations are the competing ones. However, by comparing the conditions of instability (1.5 and 1.6) it becomes obvious that as far as energy is concerned, the electron capture is more favorable. It begins with the absorption of electron, while positron decay involves emission of electron. Thus, the energy profit of K-capture is found to be $2m_e$.

In nature there exists a triad of isobars, where the mass of the intermediate radioactive atom is greater than the mass of the two extreme stable ones. If, however, the instability condition is fulfilled, the intermediate isobar can split into two ways (β^--decay, and K-capture), thus transforming into the nuclei of the extreme isobars.

The probability of a certain type of transformation is proportional to the difference of masses. The best known example of this triad is ^{40}K:

$$(12\%)^{40}_{80}\mathrm{Ar} \xleftarrow{\mathrm{K}} {}^{40}_{19}\mathrm{K} \xrightarrow{\beta^-} {}^{40}_{20}\mathrm{Ca}(88\%)$$

$${}^{40}_{18}M(\mathrm{Ar}) = 39.9623842;\ {}^{40}_{19}M(\mathrm{K}) = 39.9639998;\ {}^{40}_{20}M(\mathrm{Ca}) = 39.96215889$$

The energy that is emitted during a single β-decay varies from 0.02 MeV for ${}^3_1$H to 13.4 MeV for ${}^{12}_5$B. Since in the case of β-decay the nucleus emits two particles, the decay energy is distributed between them. Due to the static character of radioactivity, the relationship between the electron and antineutrino energies in the case of β-decay can be of any kind.

Thus, the kinetic energy of a β-particle (electron) can have an arbitrary value from 0 to E_{max} (total decay energy). For a large number of disintegrating nuclei of the same radionuclide, the distribution of β-particles by energy (*energy spectrum*) has a particular form. It was the continuous character of the β-spectra, where E_{max} is the characteristic feature of the specific nucleus, that in 1930 allowed Pauli to predict the existence of the neutrino, 25 years before it was experimentally observed.

1.2.2 Alpha Decay

Alpha decay is when a nucleus transforms into another, lighter nuclei, by emitting the alpha particle, the nucleus of helium $\left({}^4_2\mathrm{He}\right)$

$${}^A_Z\mathrm{X} \rightarrow {}^{A-4}_{Z-2}\mathrm{X} + {}^4_2\mathrm{He} \tag{1.7}$$

$$(A_2 = A_1 - 4;\ Z_2 = Z_1 - 2)$$

Alpha decay is characteristic of the heavier nuclei at the end of the periodic system, i.e. those overloaded with protons as well as with neutrons. Nuclei in the lanthanide group when A = 140 to 160 are also alpha-active.

The lightest alpha-active radionuclide is $^{145}_{28}Ce$. The energy of alpha particles that are emitted during decay varies from 4-9 MeV for heavy nuclei to 2-4.5 MeV for the lanthanides.

The energy of alpha particles depends on the half-life. According to the semi-empirical Geiger-Nuttal rule

$$\log T_{1/2} = C + \frac{D}{\sqrt{E}} \tag{1.8}$$

where C, D are constants, independent of A, though somewhat dependent on Z.

Thus,	for	$Z = 84$	$C = -50.15;$	$D = 128.8$
	for	$Z = 90$	$C = -51.94;$	$D = 139.4$

The total energy ($\Delta\varepsilon$) emitted during the alpha decay is expressed through the binding energies of the parent nucleus ($\varepsilon_{Z,A}$), daughter nucleus ($\varepsilon_{Z-2,A-4}$), and alpha particle (ε_{α}):

$$\Delta\varepsilon = \varepsilon_{Z-2,A-4} + \varepsilon_{\alpha} - \varepsilon_{Z,A}$$

When $\Delta\varepsilon > 0$ the decay is possible, whereas in the case of $\Delta\varepsilon < 0$ it is not. The binding energy of alpha particle is 28 MeV, which is 7 MeV per nucleon. For medium-weight nuclei the binding energy is approximately 8 MeV per nucleon and α-decay is not allowed. For heavy nuclei the specific binding energy is lower due to Coulomb repulsion between the protons. Approximately around $A = 190$ the specific binding energy becomes equal to 7 MeV per nucleon and alpha decay is possible. Variations of $\Delta\varepsilon$ are associated with the shell structure of the nucleus. If the parent nucleus has a completed shell, the $\Delta\varepsilon$ is lowered. If, on the other hand, it is daughter nucleus that has a completed shell, the $\Delta\varepsilon$ is increased. It is because of this kind of shell effect that the initial, rare-earth region of α-active nuclei begins with $N = 84 + 2$, and the heavy α-active nuclei are numerous starting with $Z = 84$. The α-active rare-earth nuclei are deficient in neutrons, which enhances the role of Coulomb repulsion. In these nuclei alpha decay is frequently encountered with the competing processes of K-capture and positron decay. What usually is observed is only a more rapidly occurring type of decay.

It is noteworthy that nucleon decay is less advantageous in terms of energy than α-decay. The separation of a nucleon requires nearly 6 MeV on the average. For this reason instead of four nucleons the nucleus emits one α-particle with an energy of 4-6 MeV, i.e. 4-6 times less than in the case of nucleonic decay. The energy of alpha particles ranges from 2-10 MeV. In compliance with the formula (1.8) even a small change in decay energy has marked variations in the half-life. Nuclei with E_α greater than 9 MeV split practically instantaneously, while when $E_\alpha < 4$ MeV the nuclei live so long that the decay is hard to register.

1.2.3 Gamma Radiation

During radioactive decay of any type the newly-formed nucleus may be either in its ground state, or in one of its excited levels. In the case of alpha decay, as a rule, the transitions to the excited levels are less probable than during β-decay and K-capture. Decays at excited levels cause the fine structure in the energy spectra. For instance, spectrum of ^{238}U consists of two principal lines—4.132 MeV (~22%) and 4.180 MeV (~78%); the spectrum of ^{226}Ra—4.589 MeV (5.7%) and 4.777 MeV (94.3%).

The transition of the nucleus from the excited state into its ground state is accompanied by the emission of photons of short-wave electromagnetic radiation, called *gamma radiation*. The sum of the energies of γ-photon and their corresponding charged particles is equal to the energy of transition. Moreover, we can often observe a gradual transition of the nucleus from one energy level to another (cascade transition), where

$$E^{\gamma}_{\text{transition}} = E^{\gamma}_1 + E^{\gamma}_2 + E^{\gamma}_3 + \ldots + E^{\gamma}_i$$

The energy of the γ-photon varies for different nuclei, from several keV to several MeV, which corresponds to the wavelengths from 2×10^{-8} to 5×10^{-12} cm (Table 1.1).

Table 1.1 The Energy α–, and γ-Spectra of Natural Radionuclides

Nuclide	E_α (MeV)	E_γ (MeV)
^{238}U	4.132 (~22%)	0.048
	4.180 (~78%)	—
^{226}Ra	4.589 (~5.7%)	0.188
	4.777 (94.3%)	—

The half-life of gamma-active (excited) nuclei is extremely short and is of the order of 10^{-7}-10^{-11} s. In rare cases we can find relatively long-lived excited nuclei, which are called *isomers*. An example is the nucleus $^{234}_{91}Pa$, the half-life of the isomer $^{234}_{91}Pa^{m}$ is equal to 1.175 minutes and comprises only 0.15% of the total sum of nuclei of the radionuclide, decaying with a half-life of 6.7 h $\left(^{234}_{91}Pa\right)$.

1.2.4 Spontaneous Fission

Besides α-decay the heaviest nuclei can undergo another type of transformation—spontaneous fission into two fragments comparable in mass with the emission of several neutrons. The phenomenon was discovered by K.A. Petrzhak and G.N. Flerov in 1940. Spontaneous fission was first observed for the nuclei that have mass number 232 and has a very low probability as compared with the competing alpha decay. For instance, for the artificial radionuclide $^{232}_{92}U$ the half-life for α-decay (T_α) is equal to 74 years, while for spontaneous fission (T_s) it is 8×10^{13} years.

As the nuclear mass increases, the probability of spontaneous fission becomes greater. For instance, the half-life for spontaneous fission of $^{252}_{98}Cf$ is 66 years, while for α-decay it is 2.5 years, for Kurchatovium $\left(^{260}_{104}Ku\right)$, T_s is 0.3 s. Instantaneous spontaneous fission is the main difficulty in obtaining new transuranium elements.

1.3 Radiation Effects

1.3.1 Alpha Radiation

Alpha particles are heavy charged particles. They are nuclei of helium—$^{4}_{2}He$. The energy of alpha particles emitted during alpha decay ranges from 3.15 MeV $\left(^{209}Bi\right)$ to 8.8 MeV $\left(^{212}Po\right)$ and the rate at which the radionuclides decay is determined by (1.8). Thus, ^{209}Bi has $T_{1/2} = 2 \times 10^{-17}$ years, and for ^{212}Po, $T_{1/2} = 2.9 \times 10^{-7}$ s. The energy spectrum is physical characteristic of each radionuclide and is used in analytical practice.

Passing through the matter, the alpha particle 'elbows its way' amidst the atomic electrons with its Coulomb field, thus causing either ionization, or the excitation of the matter's atoms.

In addition to this the alpha particle gradually loses its energy. The main losses of energy by α-particle $\left(-\frac{dE_\alpha}{dx}\right)$ is due to ionization and excitation of atoms. Other types of energy loss (collision with nuclei) are relatively insignificant. At the end of the path the velocity of the particle decreases due to the loss of energy whereas the residence time in the proximity of the nucleus and the probability of ionization become greater.

When the alpha particle collides with the electrons, it slightly deviates from its path because its mass is greater than those of electrons and the chaotic character of the interactions. Hence, the trajectory of alpha particles is practically a straight line. The length of the trajectory is called the *range*. The range of alpha particles in a substance R depends on the energy of the particles and the substance itself. There exists an empirical relationship between the length of the range of alpha particles in the air R_0 and their energy E_α, viz.

$$R_0 = KE_\alpha^{3/2} \quad (E_\alpha \geq 0.5\text{MeV}) \tag{1.9}$$

(for t = 15 °C, P = 760 mm Hg, K = 0.318)

Knowing the range of alpha particle in the air, it is possible to calculate its range R_x in any other substance:

$$\frac{R_x}{R_0} = \frac{\rho_{\text{air}}}{\rho_x}\sqrt{\frac{A_x}{A_{\text{air}}}} \tag{1.10}$$

where ρ is the density, and A is the mass number.

An example of the ranges of α-particles for natural series of decay (Prutkina, Shashkin, 1975) in air changes from 2.5 cm for ^{232}Th (E_α = 3.9 MeV) to 8.5 cm for ^{212}Po (E_α = 8.8 MeV). The range in aluminium can give us an idea about the range of α-particles in rocks. For E_α = 4.0 MeV it is equal to 16 μm, and for E_α = 10.0 MeV—0.9 μm. Thus, a sheet of cardboard or the walls of a glass tube will completely screen alpha radiation. Coulomb interaction between alpha particles and electrons takes place at relatively great distances (of the order of interatomic ones). Therefore, in a single collision the electron receives comparatively small energy, of the order of 10 eV.

This energy is sufficient to knock electrons out of the outer shells of atoms. Electrons in the inner shells by taking this energy can only be transferred into an excited state. Not every collision of an alpha particle with an atom will be accompanied by ionization because of the considerable scattering of the transmitted energies.

An estimate of the ionizing ability of alpha particles can be approximated as follows. It has experimentally been shown that a single ionization in the air requires E_i = 33 eV on the average by an alpha particle. By dividing the total energy of the particle E_α by E_i we get an approximate number of pairs of ions formed by the α-particle along its range. For instance, a particle with energy 6.6 MeV creates $6.6 \times 10^6/33$ = 200 000 pairs of ions on its path.

The ionizing power of alpha particles in other media is close to that which is observed in the air.

1.3.2 Beta Radiation

Beta radiation is a stream of electrons. The value E_{max} is characteristic of each β-radiator. Between the decay constant λ and E_{max} for β-radiating radionuclides there exists an approximate energy ratio:

$$\lambda = KE_{\max}^{5} \tag{1.11}$$

The value of E_{max} for natural radionuclides varies from 0.01 MeV (^3H) to 2.2 MeV(^{212}Bi). The processes that emerge during the passing of β-particles through the matter are distinguished from the interaction of heavy particles with the substance itself. During the collision of β-particle with atomic electrons, and due to its small mass, there is a marked change in momentum in each case. This means that a β-particle can considerably deviate from its initial direction. For this reason the trajectories of β-particles in the matter are not rectilinear. In practice the true range of β-particles is not determined. The maximum (extrapolated) range, or maximal absorption (R_{max}) is more useful. It is equal to the minimum thickness of the layer that wholly hinders β-particles (in g/cm²). To estimate R_{max} semi-empirical formulae are usually employed. For instance, for monoenergetic electrons the following is used:

$$R_{\max}\left[\mathrm{g/cm^2}\right] = 0.526E(\mathrm{MeV}) - 0.24 \tag{1.12}$$

Values of R_{max} are given in Table 1.2.

The continuous energy spectrum of β-radiation and its scattering by electrons results in an approximately exponential dependence of the absorption of natural

Table 1.2 Maximum Ranges of β-Particles [1.1]

E_{max}, MeV	R_{max}			E_{max}, MeV	R_{max}		
	air, m	**water, mm**	**aluminium, mm**		**air, m**	**water, mm**	**aluminium, mm**
0.01	0.00229	0.00247	0.00127	0.70	2.513	2.78	1.315
0.02	0.00773	0.00841	0.00422	0.80	2.985	3.31	1.559
0.04	0.0266	0.0290	0.0143	0.90	3.449	3.84	1.807
0.06	0.0541	0.0591	0.0289	1.0	3.936	4.38	2.059
0.08	0.0889	0.0974	0.0474	1.2	4.896	5.47	2.563
0.10	0.130	0.143	0.0693	1.4	5.868	6.56	3.070
0.20	0.407	0.448	0.214	1.6	6.821	7.66	3.574
0.30	0.763	0.841	0.400	1.8	7.781	8.75	4.074
0.40	1.168	1.29	0.611	2.0	8.732	9.84	4.593
0.50	1.601	1.77	0.837	2.2	9.683	10.90	5.074
0.60	2.050	2.27	1.070	2.4	10.611	12.00	5.593

β-radiation in matter. $I = I_0 \exp(\mu d/\rho)$; where I_0 is the original intensity of radiation; I is the radiation intensity after passing through a layer of thickness d of matter with density ρ; μ is the absorption coefficient.

From electrodynamics we know that a charged particle moving at an accelerated rate emits electromagnetic waves. For this reason the collision of a charged particle with particles of matter causes electromagnetic radiation, or *braking radiation*. The loss of energy by the particle due to braking is referred to as radiational loss. The intensity of braking radiation, or bremsstrahlung, (the amount of energy radiated per second) is determined by the formula:

$$W = \frac{2}{3}\frac{e^2 Z^2}{C^3}\left|\frac{F}{M}\right|^2 \tag{1.13}$$

where F is the force, M is the mass, Z is the nuclear charge, and C is the speed of light.

It follows, that the intensity of braking radiation in the case of a Coulomb collision with a charged centre is directly proportional to the square of the charge of the scattering centre (Z^2) and inversely so to the square of the mass of particle (M^2). Hence, the losses connected with braking are significant for electrons, but inconsequential for heavy charged particles. For instance, radiation loss for electrons is $5 \times 10^7 = (M_\alpha/m_e)^2$ times greater than for alpha particles. The bremsstrahlung for a collision between a nucleus is Z^2 times greater when it collides with the nucleus than with an electron. Therefore, radiation losses depend on the nuclei and grow in proportion to Z^2. Since the number of electrons in the nucleus is Z, the scattering of β-particles on electrons is proportional to Z, while scattering on nuclei is proportional to Z^2 (1.12). For instance, for hydrogen scattering on the nucleus and the electron is exactly the same, while for gold 99% of the scattering is on the nucleus. Therefore, protection against β-radiation is always made of a material with a low Z, while the interiors of lead measuring chambers is faced with aluminium or organic glass.

The collision between beta particles and electrons results mainly in ionization losses of energy. The relationship between radiation and ionization losses for β-particles can be approximated from the following relationships

$$\left(-dE/dx\right)_{\text{rad}} / \left(-dE/dx\right)_{\text{ioniz}} \approx \frac{ZE_\beta(\text{MeV})}{800} \tag{1.14}$$

The energy requirements for ionization for β-particles with the energies characteristic of natural radionuclides and heavy particles with similar energies are marked. Thus, the losses for protons are almost 2000 times greater than the losses for the electron of the same energy. For instance, in a photographic emulsion a proton leaves a clear trace, while an electron with the same energy does not.

1.3.3 Gamma Radiation

Gamma radiation is the short-wave electromagnetic radiation which appears during nuclear transformations, the change of energy state of a nucleus and the annihilation of particles. The wavelength λ for gamma radiation is smaller than that of the interatomic distances (10^{-8} cm). The energy of gamma radiation E_g has minimal values measuring several KeV. For natural radionuclides the maximum value E_g = 2.6 MeV: (^{208}Tl). Radionuclides are characterized by the energy spectra of gamma radiation and they are employed to distinguish elements.

Gamma radiation is absorbed by matter at the expense of electromagnetic interactions, as is the case with charged particles. However, the character of the interaction of γ-photon with matter is different: (1) γ-photon do not have any electric charge and are not subjected to Coulomb interactions; (2) the rest mass of γ-photon is equal to 0, so they cannot be retarded in the medium. They are either absorbed or deviated from their path (scattered) and leave the beam.

Range is inapplicable to γ-photon. When passing through matter the intensity of radiation decreases exponentially:

$$I = I_0 e^{-\mu d} \tag{1.15}$$

where I_0 is the initial intensity of radiation, I is the intensity after the passing through a layer of substance with thickness d; and μ is the absorption coefficient, [cm^{-1}], with μ depending on the properties of the matter and the radiation energy.

The absorption of γ-radiation by a substance takes place by three processes: photoelectric absorption (photoelectric effect), scattering on electrons (Compton effect) and the formation of an electron-positron pair.

The photoelectric effect is when a γ-photon is absorbed by an atom, as a result of which an electron is knocked out. In this case the energy of the γ-photon is transmitted to the electron (less the bond energy in an atom). The photoelectric effect is more probable if E_γ approaches the value of bond energy of the electron. The bond energy of the electron in the atom is the greater the deeper the electron shell is and the higher the atomic number Z.

Hence, the photoelectric effect in 80% of cases occurs in the K-shell and occurs the more intensely the greater Z, i.e. preponderantly on heavy nuclei (μ_f proportional to Z^5).

It is clear that the photoelectric effect is the leading mechanism of absorption for gamma quanta of low energies. The photoelectric effect has an energy threshold equal to the bond energy of a K-electron.

The Compton Effect

The Compton effect is the most significant factor of absorption for gamma photons with energies markedly exceeding the bond energies of K-electrons. The absorption coefficient for the Compton effect μ_c is proportional to Z and the number of atoms per unit volume.

Hence, Compton absorption is predominant in lighter nuclei. Scattering is defined as the interaction of a γ-photon with a free or poorly bound electron, as a result of which the γ-photon is scattered, i.e. changes its direction. Coherent and incoherent (particularly the Compton variety) scattering is possible. In coherent scattering the energy of the photon does not undergo a variation and no ionization or excitation of atoms takes place. During incoherent scattering part of the energy of γ-photon is transmitted to the atom with which it interacts.

The electron is torn away (ionization) or raised to an excited level. Compton scattering on the nuclei is negligible.

The Formation of Electron-Positron Pairs

The formation of an electron-positron pair takes place in the vicinity of a nucleus with the energies of gamma photons 1.02 MeV. The reaction cross section is proportional to Z^2. At high energies, when the reaction cross section of the photoelectric effect and Compton effect fall to 0, the formation of pairs becomes the governing factor in the absorption of electrons.

The total absorption coefficient $\mu = \mu_f + \mu_c + \mu_p$, where μ_f is the absorption coefficient at the expense of the photoelectric effect, μ_c is the Compton effect, and μ_p is the formation of pairs; μ_f is preponderant at low energies (<1 MeV) for heavy nuclei, for which reason as a protection against natural radiation the heavy elements (Pb, Fe) are used, μ_c makes the principal contribution at medium energies (several MeV) particularly on lighter nuclei. Hence, for natural gamma radiation in rocks it is scattering which is predominant; μ_p is typical of high energies.

The overall absorption coefficient for heavy elements is of minimum value within the range of energies, ~3 MeV, which must be taken into consideration when considering protection.

1.4 Radiation Dose and Units of Measuring Radioactivity

Dose is a measure of ionizing radioactivity. The action of ionizing radiation is reduced to ionization and excitation of atoms. Hence, the dose absorbed is determined by the kind of activity. The amount of radiation by the source is called *exposure dose*. The exposure dose X is determined by the charge Q of ions of the same sign emerging in the air during the complete braking of all the secondary electrons formed by a γ-photon in the mass of air dm: $X = \frac{dQ}{dm}$. The unit of exposure dose in the SI system is a coulomb per kilogram (C/kg).

A special unit is the roentgen (R), which corresponds to the dose of X-ray or γ-radiation that produces a change of 1 electrostatic unit of electricity in 1 cm^3 of air

$$1\text{C}/\text{kg} = 3.88 \times 10^3 \text{R}$$

Table 1.3 The Masses of 1 Curie of Some Natural Radioactive Elements

Radionuclide	1 Curie (mass of)	Radionuclide	1 Curie (mass of)
^{232}Th	9 t		
^{238}U	3 t		
^{235}U	465 kg	^{228}Th	1.2 mg
^{230}Th	53 g	^{224}Ra	6.2 mkg
^{231}Pa	22 g	^{210}Po	0.22 mg
^{226}Ra	1 g	^{40}K	146 kg
^{227}Ac	13.6 mg	K (natural mixture)	1230 t
^{210}Pb	18.8 mg	^{87}Rb (natural mixture)	14.9 t
			53.5 t
^{228}Ra	4.25 mg	^{144}Sm	273 t
^{90}Sr	7.3 mg	Sm (natural mixture)	1820 t
^{137}Cs	11.5 mg		
^{14}C	0.2 mg		
^{3}H	0.1 mg		

The amount of radiation per unit time is called *dose rate*: $P = \frac{dx}{dt}$; the unit of exposure dose rate in the SI system is the ampere per kilogram (A/kg). A special unit is the roentgen per hour (R/hour).

A measure of the activity of ionizing radiation is the absorbed dose D. D is the mean energy dE transmitted by radiation to a unit of matter dm, $D = \frac{dE}{dm}$. The unit of the absorbed dose in the SI system is the gray (Gy): 1 Gy, i.e. 1 J/kg. The unit that was widely used and that is still employed is the Rad, which corresponds to an absorbed energy equal to 1 R, 1 Rad = 1 c Gy = 0.01 Gy.

The biological action of radiation is connected with the damage to tissue. The latter depends on the value of the absorbed energy and on its distribution, which is determined by the linear ionization density (linear energy transfer). The higher the ionization density, the greater the biological damage. In order to take this phenomenon into account the quality coefficient Q is introduced. For the Roentgen, β- and γ-radiation, $Q = 1$. For α-radiation, $Q = 20$, and for neutron radiation $Q = 10$.

The dose that accounts for the biological efficiency of the ionizing radiation is called the equivalent dose H ($H = DQ_0$). The unit of measuring the equivalent dose is *siewert* (Sw). For γ- and β-radiation 1 Sw = 1 Gy, for α-radiation 1 Sw = 20 Gy. There is also a special unit that is used now, it is called roentgen equivalent man, or rem = 0.01 Sw.

The measure of the amount of radioactive matter is the activity A which is the number of spontaneous nuclear transformations dN per unit time dt, i.e. $A = \frac{dN}{dt}$. The activity unit is the becquerel, which is equal to 1 decay per second (Bq).

Another special and widely used unit is the curie (Ci), 1 Ci = 3.7×10^{10} Bq. One curie approximately corresponds to the activity of 1 g of ^{226}Ra. Masses corresponding to the unit of activity are inversely proportional to its decay rate (Table 1.3).

The transition from the activity of matter I (decay/s) to its quantity Q(g) is accomplished according to the formula:

$$Q = \frac{IA}{\lambda \cdot 6.02 \times 10^{23}} \tag{1.16}$$

where λ is the decay constant of the radionuclides, s^{-1}; A is the atomic mass of radionuclide; and 6.02×10^{23} is Avogadro's number.

1.5 The Basic Law of Radioactive Decay

Radioactive decay is a static phenomenon. The value which characterizes the decay of a particular radionuclide is the decay constant λ, viz. the probability of the nucleus decay per unit time.

For a large number of unstable nuclei $N\lambda$ decays will take place per unit time. It is impossible to predict the decay time for a single nucleus. In the course of time from the moment of their formation, radioactive nuclei do not 'age', and in this respect they are independent of time.

The basic law of radioactive decay is formulated in the following way: the number of atoms of radionuclide dN which disintegrate within time dt is proportional to the total number of atoms N, where λ is the proportionality coefficient (the minus sign signifying a decrease in the number of atoms with time)

$$\frac{dN}{dt} = -\lambda N \tag{1.17}$$

By integrating this equation we get the fundamental equation of radioactive decay:

$$N = N_0 e^{-\lambda t} \tag{1.18}$$

where N is the number of radionuclide atoms at present time; N_0 is the number of radionuclide atoms at initial moment; λ is the decay constant; t is the time during which the decay takes place.

The decay constant λ is specific for each radionuclide within certain range. It does not depend on the chemical compound containing the radionuclide or on the physicochemical properties of the medium (P, T). The smaller the decay constant, the longer the radionuclide.

Another constant characterizing a radionuclide is the half-life. The half-life T is the time during which the number of initially present radioactive nuclei is reduced by half. This can acquire the following expression:

$$\frac{N}{N_0} = \frac{1}{2}$$

In this case, according to the definition, $t = T$:

$$\frac{1}{2} = e^{-\lambda t}; \quad -\lambda T = \ln\frac{1}{2}$$

$$\lambda T = \ln 2 = 0.693; \quad T = \frac{0.693}{\lambda} \tag{1.19}$$

$T_{1/2}$ for each radionuclide is tabulated. Once it is known, it is not difficult to estimate how many of the initial number of atoms will remain after a time which is a multiple of the half-life: after n half-lives $\left(-\frac{1}{2}\right)^n \times N_0$ atoms will remain. Thus, in the course of $7T_{1/2}$ we shall have less than 1% of the parent atoms, and over a period of $10T_{1/2}$ their number will be nearly 0.1% of N_0.

The reciprocal of the decay constant is called the average life time of the radionuclide τ:

$$\tau = \frac{1}{\lambda} = \frac{T}{0.693} = 1.443T \tag{1.20}$$

The basic equation of radioactive decay is used to evaluate the number of atoms N_0 and N, and the decay time t, when we know two out of three unknowns (N_0, N, and t) and λ is tabulated:

$$N = N_0 e^{-\lambda t} \quad \text{— the number of atoms that remained after a certain period of time} \tag{1.21}$$

$$N_0 = N e^{+\lambda t} \quad \text{— the initial number of atoms} \tag{1.22}$$

$$t = \frac{1}{\lambda} \ln \frac{N_0}{N} \quad \text{— decay time} \tag{1.23}$$

It follows from the definition of radioactivity that one decayed parent atom corresponds to one newly-formed daughter atom. If we denote the number of daughter atoms by M, then

$$M = N_0 - N = N_0\left(1 - e^{-\lambda t}\right) = N\left(e^{\lambda t} - 1\right) \tag{1.24}$$

This equation is also widely used in calculations. If during the decay of the parent nucleus we have the formation of a chain consisting of successively transforming radionuclides, the equation has more complicated form. For two radionuclides, bound to each other in the chain (where $1 \rightarrow 2 \rightarrow \ldots$), it becomes possible to record a system of equations, where the index 1 refers to the parent radionuclide, and the index 2 to the daughter one:

$$\frac{dN_1}{dt} = -\lambda_1 N_1; \quad \frac{dN_2}{dt} = -\lambda_2 N_2 + \lambda_1 N_1 \tag{1.25}$$

The equations signify that the number of the first (parent) radionuclide constantly decreases at the expense of the decay, while that of the second (daughter) one decreases because of its own decay and is also replenished at the expense of the accumulation from the parent radionuclide.

For primary conditions:

$$N_1(0) = N_{1,0}; \quad N_2(0) = 0$$

The solution of the system will have the form:

$$N_1 = N_{1,0}e^{-\lambda t} \tag{1.26}$$

$$N_2 = \frac{N_{1,0}\lambda_1}{\lambda_2\lambda_1}\left(e^{-\lambda_1 t} - e^{-\lambda_2 t}\right) \tag{1.27}$$

The ratio between the number of two radionuclides will be equal:

$$\frac{N_2}{N_1} = \frac{\lambda_1}{\lambda_2 - \lambda_1}\left(1 - e^{(\lambda_1 - \lambda_2)t}\right) \tag{1.28}$$

If the decay of a parent radionuclide is slower than that of a daughter one ($\lambda_1 < \lambda_2$), then over a certain period of time a radioactive equilibrium will be established, when the decay rate of the daughter radionuclide equals the rate of its formation from the parent one.

In this case the value $e^{(\lambda_1 - \lambda_2)t}$ can be neglected and Eq. (1.28) will acquire the form:

$$\frac{N_2}{N_1} = \frac{\lambda_1}{\lambda_2 - \lambda_1} \tag{1.29}$$

For more long-lived parent radionuclides ($\lambda_1 << \lambda_2$) this ratio remains constant during the whole time of observation; the equilibrium is called 'secular', and Eq. (1.28) can be further simplified:

$$\frac{N_2}{N_1} = \frac{\lambda_1}{\lambda_2} \quad \text{or} \quad N_1\lambda_1 = N_2\lambda_2 = \frac{N_1}{T_1} = \frac{N_2}{T_2} \tag{1.30}$$

This relationship can be obtained from (1.25) by taking into account that secular equilibrium during the time of the experiment will practically be $\frac{dN_2}{dt} = 0$. The decrease in the number of atoms of the parent radionuclide can be disregarded, and

$$N_1\lambda_1 = N_2\lambda_2 = N_3\lambda_3 = \ldots = N_i\lambda_i \tag{1.31}$$

REFERENCES

1.1. Prutkina, M.I., Shashkin, V.L., 1975. A Handbook of Radiometric Surveying and Radiometric Analysis. Atomizdat, Moscow.

1.2. Nuclear Energetics, Man and the Environment, 1984. (Ed. by A.P. Aleksandrov). Energoatomizdat, Moscow.

2 Chemical Characteristics of Radionuclides

Most radionuclides which can be found in nature occur in small concentrations. With the exception of ^{238}U, ^{232}Th, ^{40}K, ^{147}Sm, and ^{187}Re, none ever reaches ore concentrations. All radionuclides, except ^{238}U, form ultradilute systems in natural waters. The behaviour of radionuclides in such systems ($\leq 10^{-7}$ mol/litre) in many cases differs from that of macroconcentrations of the elements whose isotopes they are, as well as from the behaviour of macroelements close in chemical properties. What may be decisive is not the chemical properties of the element, but the physicochemical state of the radionuclides (dispersion, composition of particles, chemical bonds within the substance, and the oxidation number), as well as the characteristics of the medium. The energetics of nuclear transformations also leaves its imprint on the geochemistry of radionuclides. This chapter is based on A.N. Nesmeyanov's *Radiochemistry* (1978).

2.1 The State of Radionuclides in Solutions of Small Concentrations

The state of a radionuclide in any system is the set of forms in which it exists in the solution. In ultradilute solutions radionuclides exist as ions or form colloids or pseudocolloids. In strongly acid media most radionuclides form true solutions. Reducing the acidity of the solution results in the formation of colloidal particles, and certain pH for each element marks the beginning of hydrolysis. For the group III elements colloids begin to appear at pH = 7, for groups IV and VI, the pH threshold is 3.5-5 and for group V the pH is 1-2. The isotopes of alkaline and alkaline-earth elements form only true solutions. The transition of a metal into a colloidal state as insoluble hydroxides form is described by a relationship between the pH of the transition and the concentration of the radioactive ions in the solution $[Me^{n+}]$:

$$\mathrm{pH} = (1/n)\log L_p\left[\mathrm{Me(OH)}_n\right] - \log K_{\mathrm{H_2O}} - \left(\frac{1}{n}\right)\log\left[\mathrm{Me}^{n+}\right] \qquad (2.1)$$

where n is the number of hydroxyl groups; $L_p[Me(OH)_n]$ is the solubility product of the metal hydroxide; and K_{H_2O} is the ionic product of water. The change of Me^{n+} by n orders changes the pH of transition point to the colloidal state by n units.

The presence of ions in the solution which form insoluble compounds with the radionuclide promotes the formation of true colloids (for instance, OH^- ions). Conversely, if there are substances in the solution that form soluble complexes with the radionuclide, the formation of radiocolloids is prevented. This may be accompanied by the destruction of existing colloidal particles due to a drop in the concentration of free ions in the solution, the disturbance of the equilibrium between the particle and the dissolved ions and the desorption of ions from the colloidal particle. For instance, when chloride ions are added to a solution of Po^{210}, the complex $PoCl_6^{2-}$ forms, and a colloid cannot form. The presence of electrolytes in the solution can recharge colloidal particles and change the relationship between forms of the radionuclide.

Radionuclide ions adsorbed by colloidal particles of other substances form pseudocolloids.

2.2 The Distribution of Radionuclides Between Two Phases

Radionuclide ions exist in ultradilute solutions. That is why with the substances capable of producing insoluble compounds with the given radionuclide, the solubility products will never be reached, and a solid phase will not emerge. However, if the solution contains ions of an element which is close to the radionuclide in its chemical properties and in a concentration that is adequate for precipitation, the macro- and microcomponent will coprecipitate. Coprecipitation with the formation of crystalline precipitates, where the microcomponent is distributed over the whole volume of the crystals, is called *cocrystallization.* The precipitation of the microcomponent on the surface of the solid phase of the macrocomponent (including the inner surface) is called *adsorption.* During coprecipitation, the macroelement is known as the *carrier.* If the carrier is a non-radioactive isotope of the same element as the radionuclide, this macrocomponent is called an *isotopic carrier.*

For instance, a macroconcentration of ^{86}Sr, or of a natural mixture of the isotopes of strontium, will be the isotopic carrier for the radionuclide ^{90}Sr. When a radionuclide coprecipitates with another element that is close in its chemical properties, the other element is the *non-isotopic carrier.* Examples of this are the coprecipitation of ^{90}Sr with Ca or ^{226}Ra with Ba.

Cocrystallization is described by Hahn's law, according to which small quantities of a radioactive element crystallize with the macroelement, if both form isomorphic or isodimorphic crystals with the same counterion. Khlopin's equation which expresses the cocrystallization under thermodynamic equilibrium has the form

$$\frac{-x}{y} = D\frac{a-x}{b-y} \tag{2.2}$$

where x and y are the concentrations of micro- and macrocomponents in the solid phase; a and b are the initial quantities of micro- and macrocomponents in the solution; and D is the coefficient of crystallization. If $D > 1$, then the radionuclide is accumulated in the crystals; if $D < 1$, the radionuclide is found predominantly in the solution; when $D = 1$, the ratio of micro- and macrocomponents in the precipitation and the solution are equal.

For a thermodynamically unstable systems the logarithmic equation is valid

$$\ln\left(\frac{a-x}{a}\right) = \lambda \ln\frac{b-y}{b} \tag{2.3}$$

where λ is the crystallization coefficient. In the case of logarithmic distribution the solid phase inhomogeneous.

The concentration of microcomponent grows smoothly from the centre towards the periphery when $\lambda < 1$ and decreases, if $\lambda > 1$. When equilibrium is established, the crystallization coefficient λ reaches its maximum value equal to D.

2.3 The Adsorption of Radionuclides

Adsorption precipitation was defined above as the transfer of matter from the solution to the surface of the solid phase, which is known as the adsorbent. Adsorption can take place on the solid phase with highly developed surface, e.g. finely-crystalline sediments, colloids, etc. Adsorption is very important in the hypergenic and technogenic geochemistry of radionuclides.

In the case of adsorption on polar crystals, adsorption phenomena are classified according to the location of the adsorbed ions. If the latter is in the surface layer of the crystalline lattice, viz. inside the electric double layer, the adsorption is called primary. If the adsorbed ion enters the outer part of the double layer and is in the solution at the phases interface, it is called secondary adsorption. Primary adsorption is subdivided into potential-forming and exchange classes. The former occurs for substantial concentrations of ions. Primary exchange adsorption occurs as a result of the kinetic exchange between the ions of the surface of the crystal and the solution.

Secondary adsorption is also classified into exchange and the Van der Waals adsorption. The only significant for the micro-concentrations of radionuclides are the primary and secondary exchange adsorptions.

Adsorption on ionic crystals takes place in compliance with Hahn's law, according to which the radionuclide is adsorbed on a polar crystal when the surface of the crystal has a charge that is opposite to that of the radionuclide ions. The less soluble and dissociated is the compound of the radionuclide with the oppositely charged ion of the lattice, the stronger is the adsorption.

From this definition it is clear that primary adsorption of ions is only possible if they can enter the crystalline lattice and form solid solutions with the sediments as well as yield mixed crystals. Thus, primary adsorption can be expressed as the distribution of the substance between the solution and the surface of the sediment

$$\frac{x}{1-x} = D_{\mathrm{a}} \frac{PS}{\sigma CV} \tag{2.4}$$

where x is the adsorbed fraction of the microcomponent, D_{a} is the distribution coefficient, P is the mass of the sediment, S is the specific surface of the sediment, σ is the surface occupied by 1 g-ion of the stable element, V is the volume of the solution, and C is the molar concentration of the macrocomponent (exchange ion) in the solution.

If primary exchange adsorption only takes place on the surface of the lattice, in the case of a homogeneous distribution of the microcomponent, the adsorption kinetics will be governed by monomolecular reaction law:

$$C - C_{\infty} = A + K_{\mathrm{e}}^{-\lambda t} \tag{2.5}$$

where C and C_{∞} are respectively the concentrations of macrocomponent in the solution at time t and at equilibrium; A, K_{e} and λ are constants.

In most cases, exchange adsorption is more complicated. It takes place rapidly on the surface of the solid phase. Then comes the second stage—the slow transition of the sorbed microcomponent into the solid phase—inner adsorption. In such cases the adsorption kinetics is expressed by the following equality

$$C - C_\infty = K_1 e^{-\lambda_1 t} + K_2 e^{-\lambda_2 t} \tag{2.6}$$

Secondary exchange adsorption is exponentially dependent on the charge of the ion. Indeed, experiments have shown that all things being equal, the adsorption on the negatively charged crystals NaI depends on the charge of the ion: Ra^{2+} : Ac^{3+} : Th^{4+} = 7.0 : 75.2 : 100.

The adsorption takes place in the outer layer since the cations are not isomorphic with AgI (O. Hahn, 1947).

2.4 The Isotopic Exchange

Isotopic exchange is the redistribution of isotopes of the same element in a system (inside the molecule, between molecules and phases) which does not induce any other changes. Homogeneous isotopic exchange takes place in one phase. During heterogeneous isotopic exchange isotope atoms are transferred from one phase to the other. Isotopic exchange takes place when the exchanging substances are subject to electric or thermal dissociation if reversible physicochemical processes or rapid electronic transitions may occur.

Isotopic exchange by dissociation takes place if under the same circumstances exactly the same particles (ions, molecules, and atoms) form which contain different isotopes of one particular element:

$$AX \rightleftarrows A + X$$

$$B\overset{*}{X} \rightleftarrows \overset{*}{X} + B^+$$

$$\qquad\qquad \upharpoonleft\!\downharpoonright \quad \upharpoonleft\!\downharpoonright$$

$$\qquad\qquad A\overset{*}{X} \; BX$$

An example is pairwise heterogeneous exchange between sulphur chloride and elementary sulphur:

$$S_2Cl_2 + \overset{*}{S}_8 \rightleftarrows S_2 + Cl_2 + \overset{*}{S}_2 + \overset{*}{S}_6 \rightleftarrows \overset{*}{S}_2\,Cl_2 + \overset{*}{S}_6\,S_2$$

Another is the homogeneous exchange of halide ions in solutions of metal halides and complexes with halogen atoms:

$$K_2[PtBr_6] + K\overset{*}{B}r \rightleftarrows 2K^+ + [PtBr_5]^-$$

$$+ Br^- + K^+ + \overset{*}{B}r^- \rightleftarrows K_2\left[PtBr_5\ \overset{*}{B}r_2\right] + KBr$$

Isotopic exchange by association takes place when two compounds of an element containing its different isotopes form an association (a complex, a dimer, etc.) and the reaction is reversible, i.e.

$$AX + B\overset{*}{X} \rightleftarrows AB\overset{*}{X}X \rightleftarrows A\overset{*}{X} + BX$$

An example is the reaction between hydrogen bromide and molecular bromine:

$$HBr + Br\overset{*}{B}r \rightleftarrows HBr_2\overset{*}{B}r \rightleftarrows H\overset{*}{B}r + Br_2$$

In the same way isotopic exchange can occur during all other reversible chemical processes. If different isotopes are contained in compounds of an element with different oxidation number, isotopic exchange is only possible at the expense of the transfer of an electron without any displacement of the atoms themselves. For instance, for compounds of iron with various oxidation states, an exchange is described as

$$FeCl_2 + \overset{*}{Fe}Cl_3 \rightleftarrows \overset{*}{Fe}Cl_2 + FeCl_3$$

$$\xrightarrow{\bar{e}}$$

$$Fe^{2+} + \overset{*}{Fe}{}^{3+} \rightleftarrows Fe^{3+} + \overset{*}{Fe}{}^{2+}$$

Isotopic exchange does not take place in the same manner for compounds of different classes. Binary ionic compounds readily lend themselves to exchange in conditions favourable for molecular dissociation (solutions and melts). Oxygen-containing acids and their salts easily exchange with metals (hydrogen), acid residues and oxygen. The rate of exchange is determined by the compound itself. The exchange of central atoms is hindered.

Isotopic exchange in complex compounds is easy for the ions on the outer sphere of the complex. The exchange of ligands is slower and depends on the stability constant of the compound. As a rule, the central atoms do not take part in the exchange, although very unstable complexes are exceptions.

The base readily exchanges with a metal, hydroxyl groups, or hydrogen. In organic compounds, oxygen, nitrogen, sulphur, and other atoms bonded to carbon do not take part in exchange under normal conditions. The exchange of atoms in the C—C chains is only possible for reversible chemical reactions associated with a change in the structure of the molecule, for instance, regrouping. The hydrogen of organic compounds takes part in isotopic exchange when it occurs with NH, SH, OH, etc. groups. The exchange becomes possible if the other exchange partner has a free pair of electrons:

$$\overset{*}{H}^{+} + :\underset{\cdot\cdot}{\dot{C}l}: \ldots H \rightleftarrows \overset{*}{H}:\underset{\cdot\cdot}{\dot{C}l}: + H^{+}$$

$$\overset{*}{H}^{+} + :\underset{\substack{\cdot\cdot \\ R}}{\dot{O}}: H \rightleftarrows \overset{*}{H}:\underset{\cdot\cdot}{\dot{O}}: + H^{+}$$

To describe isotopic exchange, we use the following parameters. The equilibrium constant K for the isotopic exchange for the reaction

$$A\overset{*}{X} + BX \rightleftarrows AX + B\overset{*}{X}$$

will be

$$K = \frac{[AX]\left[B\overset{*}{X}\right]}{\left[A\overset{*}{X}\right][BX]}$$

The exchange coefficient α is equal to the ratio of the fraction of the isotope molecules in the product of the reaction to the fraction in the original compound:

$$\alpha = \frac{\left[B\overset{*}{X}\right]}{[BX]} \div \frac{\left[A\overset{*}{X}\right]}{[AX]}$$

For a simple reaction (I) $\alpha = K$.

For a more complex reaction of the type $m\text{A}\overset{*}{\text{X}}_n + n\text{BX}_m \rightleftarrows m\text{AX}_n + n\text{B}\overset{*}{\text{X}}_m$

$$K = \frac{[\text{AX}_n]^m \left[\text{B}\overset{*}{\text{X}}_m\right]^n}{\left[\text{A}\overset{*}{\text{X}}_n\right]^m [\text{BX}_m]^n},$$

$$\alpha = \frac{\left[\text{B}\overset{*}{\text{X}}_m\right]}{[\text{BX}_m]} \div \frac{\left[\text{A}\overset{*}{\text{X}}_n\right]}{[\text{AX}_n]} \quad \text{where } \alpha \neq K$$

The exchange coefficient is necessary when calculating the number of stages in the isotopic exchange.

REFERENCES

2.1. Nesmeyanov, A.N., 1978. Radiochemistry, (2nd ed.). Khimiya, Moscow.
2.2. Vdovenko, V.M., 1969. The Modern Radiochemistry. Atomizdat, Moscow.

3 Primary Radionuclides. The Geochemistry of Uranium and Thorium

'There is probably no substance on Earth in which uranium, thorium, or the products of their decomposition are not present.'

'Thus, we have a clear-cut division of U and Th by geochemical processes: uranium leaves the magma as a more volatile compound and separates in hydrothermal veins (pitch-blendes) without thorium.'

V.I. Vernadsky

3.1 Primary Radionuclides

More than 230 natural radioactive nuclei are known at present. All in all they make up a relatively small fraction of the Earth's mass, i.e. only a little over 0.01% by mass. The radionuclides we find on Earth can be divided into 4 main groups by their origin.

1. The primary radionuclides are long-lived radionuclides which were formed together with the other non-radioactive nuclides of the Earth's substance in the process of overall nucleosynthesis ($T_{1/2} \geq 0.7 \times 10^9$ years).
2. The radiogenic nuclides are relatively more short-lived intermediate products of decay of uranium and thorium ($T_{1/2} \geq 2.6 \times 10^5$ years).
3. The cosmogenic are radionuclides that are constantly formed in the atmosphere and the earth's crust during natural nuclear reaction.
4. The anthropogenic radionuclides have anthropogenic origin.

In this present chapter radionuclides of the first group are discussed. The primary radionuclides have longer half-lives, which is why they remain in the Earth after nucleosynthesis. By mass they are very widely spread; as can be seen from Table 3.1 those that occur most frequently include $^{87}_{37}Rb$, $^{232}_{90}Th$, $^{40}_{19}K$ and $^{238}_{92}U$.

The average concentration of each of the above in the earth's crust exceeds 2 g/ton. Thus, they are more common Sn, Mo, W, Ta, As, Sb, I, Bi, Au, Ag, Hg, Ge, etc. According to the extent to which they are spread in the earth's crust, thorium is 47th and the 54th place is uranium. Beginning with Bi all the elements are wholly radioactive.

Table 3.1 Average Concentration of Primary Radionuclides in the Earth's Crust

Radio-nuclide	Average concentration of the sum of isotopes (acc. to Vinogradov), g/ton	Fraction of radio-active isotope in natural sum of isotopes, %	Average concentration of isotope in the earth's crust, g/ton	Half-life	Type of decay	Energy of charged particles, MeV	Energy of γ-radiation, MeV	Stable product of decay
$^{87}_{37}Rb$	150	27.8	41.6	6×10^{10}	β^-	0.275	0.394	$^{87}_{38}Sr$
$^{232}_{90}Th$	13	100	13.0	1.4×10^{10}	α	4.0	0.075	$^{208}_{82}Pb$
$^{40}_{19}K$	25000	0.0118	3.0	1.3×10^{9}	β^-	1.325	—	$^{40}_{20}Ca$
					K	—	1.459	$^{40}_{18}Ar$
$^{238}_{92}U$	2.5	99.3	2.5	4.5×10^{9}	α	4.18	0.048	$^{206}_{82}Pb$
$^{150}_{60}Nd$	37	5.6	2.1	5×10^{10}	β^-	0.011	—	$^{150}_{61}Pm$
$^{147}_{62}Sm$	8	15.07	1.2	1.05×10^{11}	α	—	—	$^{143}_{60}Nd$
$^{176}_{71}Zr$	8	2.6	0.21	3.6×10^{10}	β^-	0.215	0.180	$^{176}_{72}Hf$
$^{115}_{49}In$	0.25	95.8	0.24	6.9×10^{14}	β^-	0.63	—	$^{115}_{50}Sn$
$^{124}_{50}Sn$	2.5	6.1	0.15	1.5×10^{17}	β^-	1.5	—	$^{124}_{51}Sb$
$^{138}_{57}La$	29	0.089	0.026	1×10^{11}	K	—	1.43	$^{138}_{56}Ba$
					β^-	1.0	0.8	$^{138}_{58}Ce$
$^{235}_{92}U$	2.7	0.7	0.02	7.1×10^{8}	α	4.4	0.18	$^{207}_{82}Pb$
$^{209}_{83}Bi$	0.009	100	0.009	2.7×10^{17}	α	3.15	—	$^{205}_{81}Tl$
$^{187}_{75}Re$	0.0007	62.9	0.0004	4×10^{12}	β^-	0.04	—	$^{187}_{76}Os$
$^{130}_{52}Te$	0.0012	34.1	0.0003	1.4×10^{21}	β^-	1.0	—	$^{130}_{53}I$
$^{144}_{60}Nd$	37	23.37	8.85	2×10^{15}	α	—	—	$^{140}_{58}Ce$
$^{123}_{52}Te$	0.0012	0.88	1×10^{-5}	1.2×10^{13}	K	—	—	$^{123}_{51}Sb$
$^{180}_{74}W$	1.3	0.126	0.0016	2.2×10^{17}	α	3.2	—	$^{176}_{72}Hf$

The shortest half-life of the radionuclides in this group is expressed as $^{235}_{92}U : T_{1/2} = 7.1 \times 10^8$ years. This illustrates that all radioactive nuclei with half-lives shorter than 10^8 years which may once have existed in the Earth have practically all decayed. The quantity of the remaining radionuclides has fallen considerably since the formation of the Earth's matter: approximately 12 times for ^{40}K and 30 times for ^{235}U.

3.2 The Chemical Properties and Minerals of Uranium and Thorium

Uranium and thorium are the most widespread and applied radioactive elements. They are members of the actinides family, which also contains actinium, protactinium, and plutonium. This is why their chemistries are to a great extent similar. The lanthanide family (5*f* elements) is chemically analogous to the actinides (4*f* elements). Affinity with the lanthanides is important in the geochemistry of uranium and thorium. In size the ionic radii of U^{4+} is closer to the 'heavy' lanthanides (the yttrium group), and Th^{4+} to the 'lighter' ones (the cerium group), see Table 3.2. Because the actinides and lanthanides are so similar, all rare-earth minerals contain variable quantities of uranium and thorium as isomorphic impurities. Moreover, U is preponderant in the 'yttrium' minerals, while Th in the 'cerium' minerals. For instance, there is a heterovalent isomorphism:

$$^{3+}Ce + PO_4^{3-} \rightarrow Th^{4+} + SiO_4^{4-} \text{ (in monozite)}$$

$$^{3+}Y + PO_4^{3-} \rightarrow U^{4+} + SiO_4^{4-} \text{ (in xenotime)}$$

At the same time the similarity of the ionic radii (and equality of charge) between U^{4+} and Th^{4+} results in the identity of their chemical properties. In endogenic minerals, where uranium is tetravalent, U and Th often give rise to isostructural series, as in the following cases:

$$UO_2 \text{ (uraninite)} \rightleftarrows ThO_2 \text{ (thorianite)}$$

$$USiO_4 \text{(coffinite)} \rightleftarrows ThSiO_4 \text{(thorite)}$$

Chemical Properties of Thorium. In natural compounds Th has only one oxidation state (IV). Compounds with Th in a lower degree of oxidation (II and III) are artificial. They are not very stable. The element oxidizes to ThO_2

Table 3.2 Ionic Radii of Actinides and Lanthanides (Ahrens)

Element	Group	Valence			
		3^+	4^+	5^+	6^+
Ac	Actinides	1.41*	—	—	—
Th		(1.08)*	1.02	—	—
Pa		(1.05)*	0.98	0.80	—
U		1.03*	0.97	0.87*	0.80
La	Lanthanides	1.14	—	—	—
Ce		1.07	0.94	—	—
Pr		1.06	0.92	—	—
Nd		1.04	—	—	—
Sm		1.00	—	—	—
Eu		0.98	—	—	—
Gd		0.97	—	—	—
Tb		0.93	0.81	—	—
Du		0.92	—	—	—
Ho		0.91	—	—	—
Er		0.89	—	—	—
Tm		0.87	—	—	—
Yb		0.86	—	—	—
Lu		0.85	—	—	—
Y		0.92			

* According to Zakhariasen

dioxide, which is its most important compound. ThO_2 has a very high melting point (3222 °C) and does not readily dissolve. As distinct from TiO_2 and ZrO_2, ThO_2 dissolves rather easily in sulphuric. Thorium hydroxide $Th(OH)_4$ precipitates from the solution when pH ≥ 3.5, and is soluble in acid. In the presence of alkali carbonates and salts of organic acids it forms soluble complex compounds.

The thorium halides, with the exception of the fluoride, are readily soluble and volatile at high temperatures. They are readily hydrolized by steam to form oxyhalides ($ThOCl_2$, $ThOBr_2$).

Thorium sulphate (dehydrated and in the form of crystal hydrate) is soluble, though its solubility is lower than those of the analogous compounds of the cerium lanthanides. The nitrate is the most soluble thorium compound and it forms several crystal hydrates.

The fluoride $ThF_4 \cdot 4H_2O$; iodate $Th(IO_3)_4$, oxycarbonate $ThOCO_3 \cdot H_2O$; oxalate $Th(C_2O_4)_2 \cdot 6H_2O$, sulphite $Th(SO_3)_2 \cdot 2H_2O$, phosphates, chromates, and molybdates are all practically insoluble.

Owing to its large charge and relatively small ionic radius, thorium tends to form complexes.

In solutions where pH $\leq$ 3 thorium exists as the ion Th^{4+}. When the pH is increased, the ion is hydrolyzed:

$$Th^{4+} + 2H_2O \rightleftarrows Th(OH)_2^{2+} + 2H^+$$

$$2Th^{4+} + 2H_2O \rightleftarrows Th_2(OH)_2^{6+} + 2H^+$$

when pH > 3.5, microconcentrations of thorium form a colloidal $Th(OH)_4$ in the solution.

With monobasic acids thorium forms complexes of the type: ThA^{3+}, ThA_2^{2+}, ThA^+ (A is the anion of the acid).

Analogous complex ions emerge with sulphuric and phosphoric acids. With nitric acid we get the complex $Me[Th(NO_3)_5]$, and with hydrofluoric acid Me $[ThF]_5$, and $Me_2[ThF]_6$. With an excess of carbonate ions, we get $Me_6[Th(CO_3)_5]$. Thorium forms many soluble complex compounds with the salts of organic acids (tartrates, citrates, etc.). With ammonium oxalate and the alkaline metals, soluble complexes of the type of $Me_4Th(C_2O_4)_4$ are formed [3.1, 3.2].

Thorium Minerals. There are more than a hundred natural thorium-containing minerals. In the minerals thorium is present as a tetravalent cation with a coordination number of 8. Thorium is a typical lithophile element. It has a high affinity for oxygen and so is found exclusively in oxygen compounds (oxides, silicates, phosphates, carbonates, and fluorocarbonates). No sulphides, selenides, or halides of thorium have been discovered in nature as yet.

The isomorphic and paragenetic bond of thorium with rare-earth elements is also common, particularly with the cerium series and tetravalent uranium. There are only a few thorium minerals proper.

Thorianite (Th, U)O_2 contains 45-93% of ThO_2; it is isomorphous with uranite, thus forming with it complex transitional species. It usually includes isomorphous impurities of cerium group elements (up to 13%).

The cerium variety of the mineral is cerianite Ce(Th)O_2 (about 5% of ThO_2).

Thorite $ThSiO_4$ contains from 50 to 77% of ThO_2. It usually contains admixtures of uranium, rare-earth elements of cerium group and iron impurities.

Thorite includes uranothorite and ferrithorite, which differ in their concentrations of uranium and iron, as varieties. Thorite with a large water content is called orangite. A morphological variety of thorium (the monoclinic one) is known as huttonite.

The hydrated thorium silicate is thorogummite (from 18 to 50% of thorium). It is isomorphous with thorite and is a product of its transformation. Cheralite (Th, Ca, Cl) (PO_4, SiO_4) contains up to 32% of ThO_2 and 4% of U_3O_8. Cheralite is an intermediate member in the series starting with $ThSiO_4$ (thorite) and ending with $CePO_4$ (monacite).

Monacite is the most important industrial mineral source of thorium. It usually contains 2.5-12% (up to 28%) of ThO_2 and of a few tenths of one percent of UO_2. Sometimes monacites without thorium and uranium can be found.

Considerable amounts of thorium (up to 13%) exist as complex oxides of titanium, tantalum, and niobium (pyrochlore, aeschynite, euxenite, samarskite, absite, and others).

Chemical Properties of Uranium. In nature uranium exists in two oxidation states, viz. +4 and +6. Under laboratory conditions unstable U(III) and U(V) can be obtained [3.1, 3.2].

Uranium (IV) is present in compounds in the form of U^{4+} ion. In its chemical properties it is very close to Th(IV), Y(III), and the heavy lanthanides of the yttrium group. In chemical terms, the U^{4+} ion is a weak base. It only exists in very acidic solutions and tends to hydrolyze to form $U(OH)_4$ when the acidity is low. The hydroxide $U(OH)_4$ is poorly soluble in water, though it completely dissolves in acids. Uranium dioxide UO_2 does not react with water up to 300 °C and is not soluble in HCl. However, it dissolves quite well in HNO_3 and mixtures of acids. The most soluble compounds of U(IV) in water are UCl_4 and $U(SO_4)_2$. Thus U(IV) is stable in the acid sulphate, chloride and nitrate solutions. U(IV) silicates are soluble in very acidic media. U(IV) tends to form complexes in which its coordination number is 8. Carbonate, sulphate, oxalate and other complexes of U^{4+} exist. Most are not stable and their stability drops with increasing temperature. For instance, in carbonate solutions, U(IV) forms a stable complex of $U(CO_3)_5^{6-}$ carbonate in excess.

U(VI) is the most stable oxidation state of uranium when air is available. A state of 6 is high even for such a large cation as U^{6+}. U^{6+} is not energetically stable and in water solutions is instantly hydrolyzed to form a complex cation UO_2^{2+}.

For example: $UF_6 + 2H_2O \rightarrow UO_2F_2 + 4HF$.

The uranyl ion is the largest natural cation with a diameter of 2.64 × 6.84 Å [3.3]. Correspondingly the UO_2^{2+} ion cannot isomorphously replace other cations in chemical compounds and easily forms its own minerals in nature.

UO_3 is a very active chemical compound. With acids, it yields solutions of uranyl salts. According to Makarov, UO_3 is found in minerals (uranites and pitchblendes) in the form of uranyl groups, where O^{2-} ions penetrate into the voids to compensate the positive charge $UO_2^{2+} \times O^{2-} = UO_3$. Compounds of U(VI) dissolve readily and are stable in aqueous solutions. The most soluble are the uranyl nitrates, as are the sulphates and oxyhalides (UO_2Cl_2 and UO_2F_2). UF_6 and UCl_6 are volatile, but in the presence of steam they tend to be easily hydrolyzed into oxyhalides. Many uranyl salts are soluble in organic acids. The most significant, insoluble U(VI) compounds are the phosphates, arsenates, and vanadates.

U(VI) readily form complexes and this is important in its geochemistry. In all these compounds uranium is found in the form of uranyl ion, the coordination numbers of which are 4 and 6. The complex formation of U(VI) with Cl^-, Br^- and NO_3^- is very slow. The most significant compounds in nature are the carbonates, sulphates, fluorides, phosphates, and hydroxyl complexes. The complexes of the uranyl ion with hydroxyl ions and water are formed by the step replacement of water groups in aqua-ion of uranyl $[UO_2(H_2O)_6]^{2+}$ and have the form

$$[UO_2(OH)_n(H_2O)_{6-n}]^{2-n} \quad (n = \text{from } 0 \text{ to } 6)$$

Carbonate complexes form families of compounds, of which the more stable in aqueous solutions are $[UO_2(CO_3)_3]^{4-}$ and $[UO_2(CO_3)_2(H_2O)_2]^{2-}$. The former occurs in solution with an excess of CO_3^{2-} and in the case of dissolution becomes transferred into the latter. The next step is the formation of weakly soluble uranyl carbonate UO_2CO_3 (Fig. 3.1). Fluoride complexes are only formed with high concentrations of fluorine. For example, at 25 °C the threshold concentration of fluorine should be 0.5 mg/l [3.4]. The sulphate complex compounds of uranyl have much in common with the carbonates as far as structure is concerned. However, in terms of strength they are weaker than both the carbonates and fluorides. They are only characteristic of the acid medium with pH = 2-4 (Fig. 3.2). An important group of complex compounds is formed by uranyl ion with the organic acids (oxalic, acetic, citric, the group of humus acids, etc.).

Redox reactions are important for the geochemistry of uranium. The high stability of U(VI) in aqueous solutions and the low stability of U(IV), ensures the dispersion of uranium in some of the physicochemical conditions and its accumulation in others, which in a number of cases results in the formation

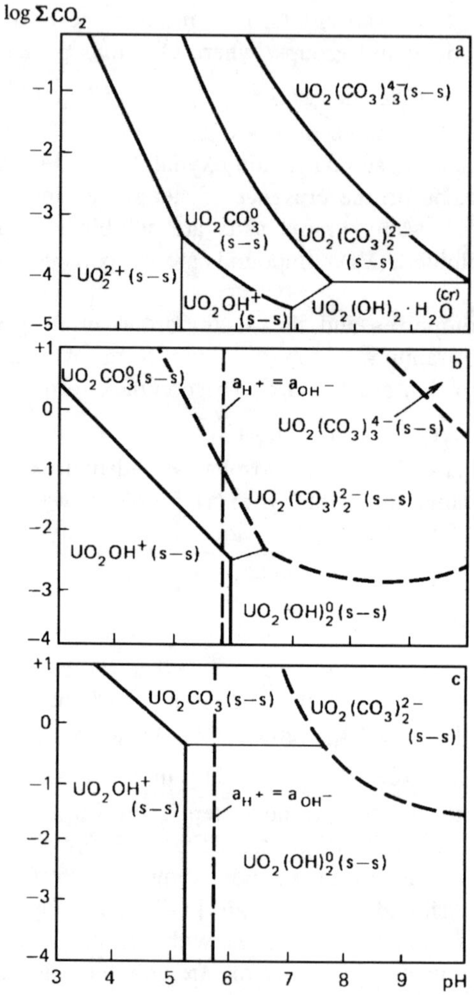

Fig. 3.1. Fields of preponderance of various ions of uranile in coordinates CO_2-pH at 25 °C (a), 150 °C (b), and 300 °C (c) (according to G.B. Naumov, 1978)

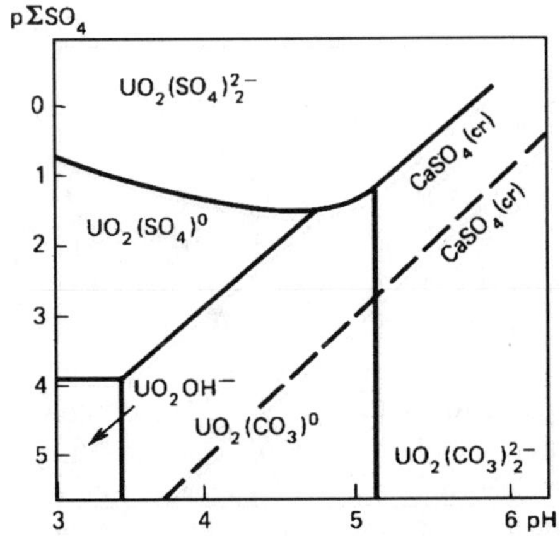

Fig. 3.2. The relation of stability fields of sulphate and hydrocarbonaceous complex ions of uranile (SCO_2 = 1 mol/kg H_2O, t = 150 °C) (according to G.B. Naumov, 1978)

of deposits. The system $U^{4+} \rightleftarrows UO_2^{2+}$ is unsufficiently reversible. With the absence of complex-forming factors and low natural concentrations of uranium in aqueous solutions this reaction is:

$$U^{4+} + 2H_2O \rightleftarrows UO_2^{2+} + 4H^+ + 2e^-$$

The redox reaction of one element is always associated with that of another element. Oxidation can involve oxygen ($O_2 \rightarrow O_2^- + e^-$), any other anion generator, and, finally, an ion getting a higher redox potential.

The most significant redox system for the geochemistry of uranium is the association of uranium and iron. According to Shcherbina [3.5] the reaction in an acidic sulphate medium is

(1) $Fe^{2+} \rightarrow Fe^{3+} + e^-$ $(E_0 = +0.771V)$

will be oxidizing as compared to the reaction:

$$(2)\ U^{4+} + 2H_2O \rightarrow UO_2^{2+} + 4H + 2e^- \quad (E_0 = +0.407V)$$

The sum is taken according to the scheme:

$$2Fe^{3} + U^{4+} + 2H_2O \rightarrow 2Fe^{2+} + UO_2^{2+} + 4H \quad (E_0 = 0.567V)$$
$$E_0 = (2 \times 0.771 - 0.407)/2 = 0.567$$

At the same time the oxidation of uranium requires the reduction of iron.

During a transition from an acid to an alkaline medium, the potentials change and become lower. Since an unequal number of electrons take part in the reactions, the elements in the redox series often change places.

For example, in an alkaline medium the uranium-iron system will have the form:

$$(1)\ Fe(OH)_2 + OH^1 \rightarrow Fe(OH)_3 + e^- \quad (E_0 = -0.56V)$$

$$(2)\ U(OH)_4 + 2(OH^1) \rightarrow UO_2(OH)_2 + 2H_2O + 2e \quad (E_0 = -0.49V)$$

Since $E_0^U > E_0^{Fe}$, reaction (2) will be more electropositive than (1) and uranium is the oxidizer with respect to iron. Thus, the reaction

$$2Fe^{2+} + U^{6+} \underset{acid}{\overset{alk}{\rightleftarrows}} 2Fe^{3+} + U^{4+}$$

in acid media goes from right to left, and from left to right in alkali media.

In the reduction reactions of uranium organic compounds are important. The reducing agent is either the organic compound, which is oxidized, or H_2S or Fe(II), which are reduced by it. The composition of the organic compound is important too. Saturated hydrocarbons are practically never reducing agents. Unsaturated hydrocarbons with —C=C— double bonds have clearer reducing properties. They are even more pronounced in oxyacids. For instance, during the decomposition of a cellulose groups of sugars form, producing a powerful reducing agent, such as ascorbic acid. Reducing properties decrease with increasing carbonification of plant remains, viz., brown coal > anthracite > schungite > graphite.

Uranium Minerals. True uranium minerals are divided into primary minerals of uraniumU(IV) and secondary uraniumU(VI) minerals.

U(IV) minerals contain uranium in the form of the U^{4+} ion. The basic minerals are dehydrated oxides. They vary in composition from almost pure UO_2 ($UO_{2,17}$) to practically pure UO_3 ($UO_{2,92}$). The uranite group includes crystalline forms with isomorphous impurities of Th and rare-earth elements. The varieties containing ThO_2 impurities (up to 15%) are called bröggerite, and the varieties comprising a substantial impurities of rare-earth elements (up to 10%) are called cleveite. In the isostructural series of uranite-thorianite, the latter includes minerals with the content of ThO_2 of more than 58%. The pitchblende group includes oxides of uranium without thorium, which have a cryptocrystalline structure and colloform separation. Powder-like varieties are known as uranium blacks.

Uranium silicates, coffinite and nenadkevite, are more rare.

Coffinite ($USiO_4$) is isostructural with thorite ($ThSiO_4$) and zircon ($ZrSiO_4$). In coffinite part of the silicate tetrahedra is replaced by the $(OH)_4$ group, as a result of which it is attributed the more complex formula: $U(SiO_4)_{1-x}$ $(OH)_{4x}$. Nenadkevite is a complex mineral whose formula is assumed to be

$$(U^{4+},\ Y^{3+},\ Ce^{3+},\ Th^{4+}) \times U^{6+}(Ca^{2+},\ Mg^{2+},\ Pb^{2+})$$

$$\times\ (SiO_4)_2 \times (OH)_4 \times n(H_2O)$$

Complex oxides of titanium, tantalum, niobium and rare earths usually contain isomorphic impurities of U and Th. Economically the most important is uranium titanite, known as brannerite (UTi_2O_6), and the complex titanite of iron or davidite, which comprises up to 20% of U, viz., $(Fe^{2+}, Ce^{3+}, U^{4+})_2 \times (Ti, Fe^{3+}, V, Cr)O_{12}$. The other minerals in the group containing uranium, though less of it, include betafite, samarskite, fergusonite, hatchettolite, and euxenite. Rare earth phosphates, primarily xenotime, sometimes monacite, also contain noticeable isomorphous impurities of uranium.

All the primary minerals of uranium which contain U(IV) are dark in colour.

U(VI) minerals correspond to the oxidation zone and have the bright colours (yellow, green, orange, and white) typical of the uranyl salts. The basic structural unit of these minerals is UO_2^{2+}. Many of them luminesce in ultraviolet light. The uranyl ion has a large radius and a relatively low charge. This is related to the moderate energy of the crystalline lattice, hardness, and solubility. Practically all uranyl minerals dissolve in mineral acids, and only some in water.

The uranyl carbonates, sulphates, vanadates, arsenates, phosphates, and silicates are common. They are all double salts or trisalts. The hydrated basic phosphates, arsenates and vanadates of UO_2^{2+}, Ca^{2+}, Mg^{2+}, Ba^{2+}, Cu^{2+}, Pb^{2+} comprise the uran-mica group. They range in colour from yellow to green, join perfectly and lustre, allows them to form micaceous crystals.

The following minerals are examples

autunite	$Ca(UO_2)_2 \cdot (PO_4)_2 \cdot 10H_2O$;
metatorbernite	$[Cu(H_2O)_4] \cdot [(UO_2)_2(PO_4)_2] \cdot 4H_2O$;
metazeunerite	$[Cu(H_2O)_4] \cdot [(UO_2)_2(AsO_4)_2] \cdot 4H_2O$;
carnotite	$K_2(UO_2)_2(UO_4)_2 \cdot (0\text{-}3)H_2O$;
bayleyite	$Mg_2(UO_2)(CO_3)_3 \cdot 18H_2O$;
schroeckingerite	$Na_2Ca_3(UO_2) \cdot (CO_3)_3(SO_4) \cdot 10H_2O$;
uranopilite	$(UO_2)_6(SO_4)(OH)_{10} \cdot 12H_2O$;
uranophane	$Ca(UO_2)_2(Si_2O_7) \cdot 7H_2O$

3.3 The Geochemistry of Uranium and Thorium in a Magma Process

3.3.1 The Distribution of Uranium and Thorium in Igneous Rocks

Vinogradov estimated the average content of uranium in the earth's crust to be 2.5 ppm and that of thorium 1.3 ppm. These elements are known to be dispersed though not altogether rare.

The average contents of uranium and thorium in magma rocks are given in Tables 3.3 and 3.4. However, these contents vary markedly within the same petrochemical type of rocks depending on the formation to which they belong. The concentration of radioactive elements within the same magmatic formation will depend on whether a particular rock belongs to a certain series according to alkaline properties, the concentration gradually increasing from the tholeitic series through the alkali-lime towards the alkaline series.

According to [3.6], a characteristic of magmatic formations in the geosynclinal stage of the development of mobile belts is the low initial content of uranium and thorium in ultrabasic, basic and moderately acidic magmas, the similar behaviour of uranium and thorium, and the poorly expressed accumulation of these elements in the final products of magmatic differentiation. The low concentrations can be explained by their moderate concentrations in the original mantle source of these rocks.

For the formations in the inversion stage in the development of mobile belts, when large multiphase granite intrusions are formed, the content of uranium and thorium is normal and close to the clark. At post-inversion (orogenic) stage,

Table 3.3 Average Content of U and Th in Continental Igneous Rocks (Smyslov, 1973)

Rocks	U, ppm	Th, ppm	Th/U	Series
		Intrusive		
Peridotites, pyroxenites	0.03	0.08	2.7	Calc-alkalic
Gabbro, diabases	0.6	1.8	3.0	
Diorites	1.8	6.0	3.3	
Quartz diorites, granodiorites	2.1	8.3	4.0	
Plagiogranites (K < Na)	2.7	9.6	4.0	
Granites (K ≈ Na)	4.5	18.0	4.0	
Alkaline granites (K > Na)	6	25	4-5	
		Effusive		
Diabases, basalts	0.7	2.3	3.2	
Andesites	1.2	4.0	3.3	
Dacites	2.5	10.0	4.0	
Liparites	4.5-5.3	11-28	2-6	
		Intrusive		
Miaskitic-nepheline syenites	4.1	7.2	1.8	Alkaline
Agpaitic-nepheline syenites	10.3	28.3	2.8	
		Effusive		
Trachybasalts	2.4	8.0	3.3	
Trachyandesites	3.5	15.0	4.2	
Trachytes, trachyliparites	3-8	30-50	4-10	

Table 3.4 Average Content of U, Th (ppm) and K (%) in Basic Oceanic Rocks

Type of rock	Th	U	K	Th/U	(K/U)·10^{-4}
		Mid-oceanic ridges, basalts:			
Depleted**	0.020	0.010	1060	2.0	1.06
Enriched**	0.55	0.18	1920	3.1	1.06
		Oceanic islands, basalts:			
Tholeiites*	0.72	0.20	2600	3.6	1.3
Alkaline*	2.76	0.68	10 000	4.02	1.47
Alkaline**	3.4	1.1	9600	3.1	0.87
		Island arcs, basalts:			
Tholeiites**	0.25	0.10	3240	2.5	3.2
Calc-alkalic **	1.1	0.36	8640	3.2	2.4
		'Undepleted mantle'			
Normalized value**	0.050	0.013	120	3.85	0.92

* According to Kravchenko (1983), [3.7].
** According to Sun (1980), [3.8].

lithophile elements accumulate. A general geochemical characteristic of volcanic and volcanic-intrusive formations of the post-inversion stage in the development of mobile belts is the much higher contents of uranium, thorium and potassium than geosynclinal effusions, which are analogous in terms of their silicic properties. Within each series of rocks (tholeiitic, calc-alkalic, and alkaline) there is a regular increase in the content of uranium and thorium in the series of rocks: the ultrabasic < basic < intermediate < acidic, which is proportional to the increase in SiO_2 (Fig. 3.3). We can also find a positive correlation between uranium, thorium, and potassium (Fig. 3.4). This can be exemplified by the basalts of the Great Fractured Tolbachik eruption of 1975-76:

Rocks	U, ppm	Th, ppm	Th/U	K%
Magnesian basalts of moderate alkalinity (Northern outburst)	0.4	0.6	1.4	0.77
Subalkaline basalts (Southern outburst)	1.2	1.3	1.1	1.5

There is no correlation between uranium, thorium and potassium in the case of ultrabasic alkaline rocks. There is a clear relation between the concentration of radioactive elements and the agpaitic coefficient:

$$K_\alpha = \frac{Na_2O + K_2O}{Al_2O_3}$$

As a rule miascitic nepheline syenites with $K_\alpha << 1$ have less radioactivity than agpaitic variants with $K_\alpha > 1$ (Table 3.3), [3.9].

All the forms in which uranium and thorium exist in magmatic rocks can be reduced to 3 groups: (1) true minerals; (2) isomorphous inclusions into accessory minerals; and (3) dispersed forms. The relationships between the individual forms depend on the composition of the rocks and their crystallization conditions.

The dispersed forms of uranium and thorium are dominant in ultrabasic, basic, and intermediate rocks of the calc-alkalic and tholeiitic series.

Radioactive elements in intrusive rocks are dispersed in the lattices of rock-forming minerals. When the melt is cooled rapidly in the process of formation of volcanic rocks, the uranium and thorium contained in the residual melt become concentrated in a glass as a solid solution. The grains of minerals comprise 100-1000 times less uranium and thorium (Fig. 3.5), [3.10]. Subsequent recrystallization of volcanic rocks results in the redistribution of the radioactive elements. Most of the dispersed uranium and, to a lesser degree, of thorium is transformed under these circumstances into sorptive forms on the faces of mineral grains and in microfissures. These forms are easily recovered by carbonate solutions and have been called mobile.

In rocks with a high concentration of thorium and rare-earth elements there are more of the less mobile isomorphous forms. True minerals only occur in acidic intrusive rocks with a low concentration of rare earths. In addition, the

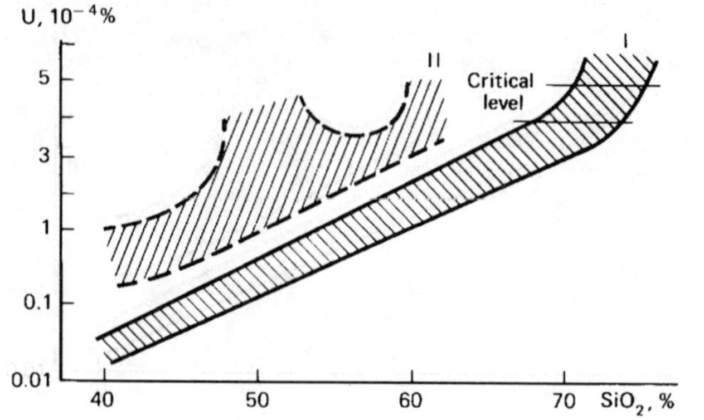

Fig. 3.3. Dependence of U content on SiO_2 in different series of igneous rocks:

I—calc-alkalic; II—alkaline (according to Smyslov, 1974)

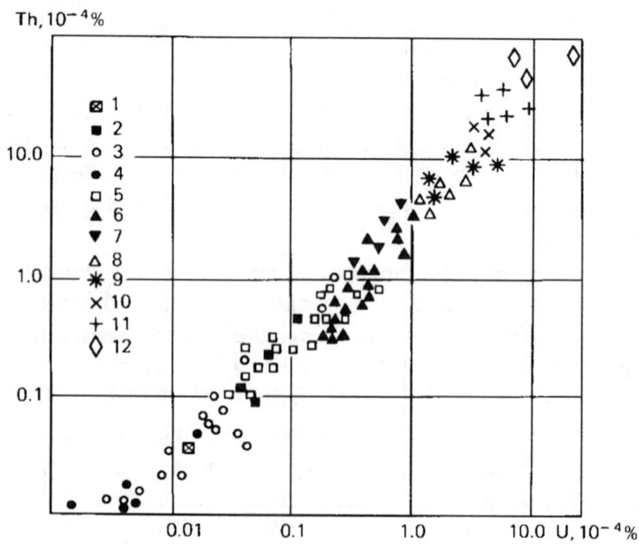

Fig. 3.4. The relation of uranium and thorium in intrusive rocks:

1—chondrite; *2*—garnet periodotites; *3*—spinel peridotite; *4*—Alpine hyperbasite; *5*—eclogite; *6*—continentaltholeiite; *7*—gabbro; *8*—diorite and quartz diorite; *9*—plagiogranite; *10*—granodiorite; *11*—alkaline granite; *12*—granite (according to Smyslov, 1983)

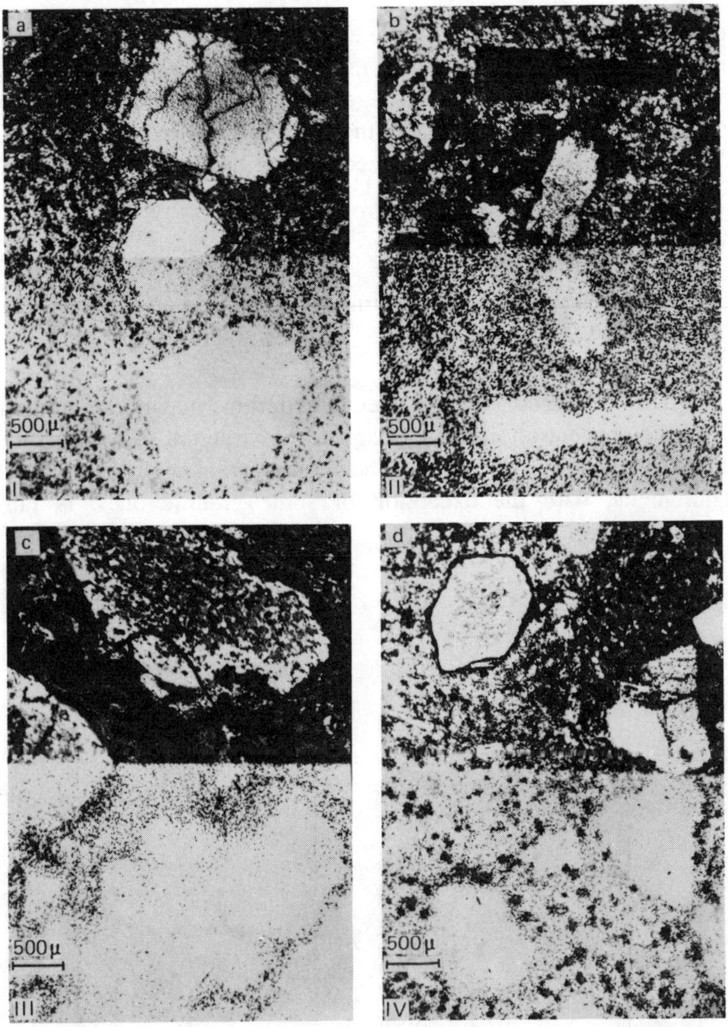

Fig. 3.5. Distribution of uranium in Quaternary volcanic rocks. Microphotographs of sections (*a*) and their correlated *f*-radiographies upon lavsan detectors (*b*) (according to Titayeva, Berman, Ishbulatov et al. 1987):

ç—plagioclase and pyroxene grains in trachyte of Baikal rift zone ($D_U^{Pl} = 0.005$; $D_U^{Px} = 0.006$); II—plagioclase grains in trachyte of Baikal rift zone ($D_U^{Pl} = 0.005$); III—plagioclase and pyroxene grains in the basalt of Klyuchevskoi volcano of 1946 eruption, Kamchatka ($D_U^{Pl} = 0.057$; $D_U^{Px} = 0.057$); IV—plagioclase and pyroxene grains in subalkaline basalt of the Grand Fissured Tolbachik eruption of 1976, Kamchatka ($D_U^{Pl} = 0.007$; $D_U^{Px} = 0.004$)

inclusions of true uranium minerals are extremely small and cannot be observed by optical microscope. In most cases, they are the thorium variants of uranite, uranothorite, and sometimes braunerite, davidite-coeffinite, betafite, and others.

Magma deposits of uranium and thorium do not exist as such. Granitic intrusions, even with relatively high concentrations of radioactive elements, have no economic value today.

3.3.2 The Behaviour of Uranium and Thorium in a Magmatic Process

The behaviour of an element during magmatic crystallization depends on its concentration in the magma, the structure of the crystals it can enter, and its own crystal chemistry. The concentration of uranium and thorium in magma melts, with the exception of some granitic ones, is fairly low

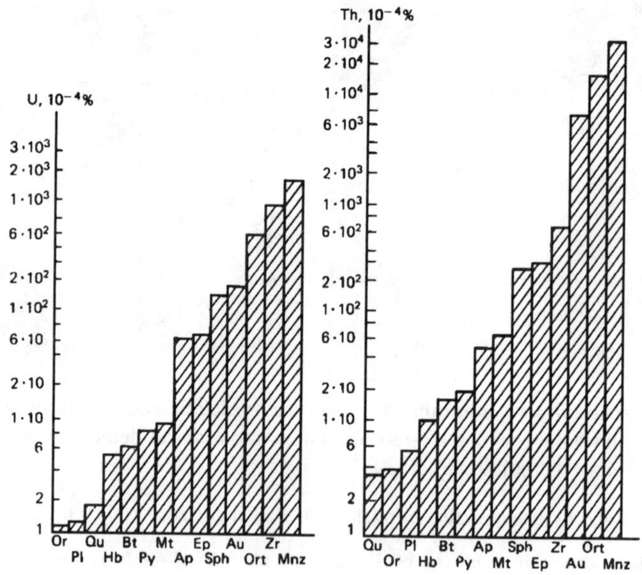

Fig. 3.6. The content of uranium and thorium in rock-forming and accessory minerals of granitoids (according to Smyslov, 1974)

(n ppm), which does not make it possible for them to form true minerals. Hence, in the process of magmatic crystallization, uranium and thorium will to some extent be incorporated into the minerals of other elements. In the reducing conditions of magma chambers uranium exists as U^{4+} and is the chemical analogue of Th^{4+}. The large charge of U^{4+} and Th^{4+} and relatively large ionic radius do not correspond to any petrogenic elements, and do not allow them to enter the lattices of rock-forming minerals isomorphically. The slow process of crystallization leads to the purification of early mineral phases from impurities and the removal of uranium and thorium into the residual melt.

In fresh volcanic rocks where the melt is rapidly quenched, practically all the radioactive elements are concentrated in glass (Fig. 3.5). In the intrusive process both uranium and thorium differentiate and the residue of the melt may be enriched with radioactive elements by hundreds and thousands of times. In usual natural melts with clark concentrations of U and Th they are mostly present in the accessory minerals. The holocrystalline intrusive rocks may contain at least 50% of the uranium and even more of the thorium (Figs 3.6, 3.7).

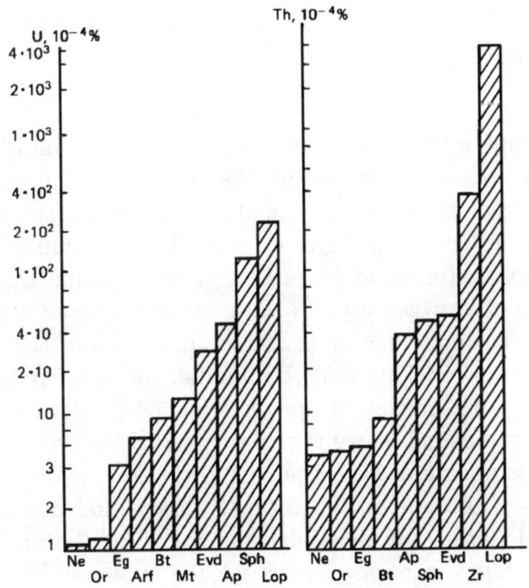

Fig. 3.7. The content of uranium and thorium in rock-forming and accessory minerals of nepheline syenites (according to Smyslov, 1974)

There are data indicating that in the late-magmatic stage of development of intrusions, uranium can leave the magma chamber. During this period, the partial pressure of oxygen rises, which promotes the oxidation of uranium and its mobility. The Th/U ratio will increase as a result of the removal of the uranium. However, if a melt has a large enough concentration of elements close in crystallographic properties to U(IV) (e.g. rare-earth elements, Th, Zr) to form their own minerals, uranium will remain firmly bound as an isomorphous impurity. Only if the concentration of such an element is insignificant will most of the uranium remain in a dispersed state and be easily removable by the outflowing fluid.

It is necessary to bear in mind that the main source of uranium extracted by postmagmatic fluids are upper crustal rocks, rather than mantle rocks. Even very alkaline or acidic mantle rocks contain small concentrations of uranium. When there is crustal magmatism of an alkaline-earth or alkaline profile tangible quantities of uranium are only separated when there is a prolonged crystallization of the magma in an enclosed space. The prerequisite is a high oxidizing potential of the medium, which also only occurs in crystal melts.

3.4 The Geochemistry of Uranium and Thorium in a Pegmatitic Process

The residual melt formed by the fractional crystallization of the magma is generally a silicate liquid, rich in alkalies, aluminium, and containing water as well as other volatile substances, and also rare elements which cannot enter into the crystal lattices of the igneous rock. The pressure of the volatile components is the driving force, which pushes them along the fractured zones in the enclosing rocks or between crystallized parts of the same intrusive body.

With gradually decreasing temperature and the increasing role of the volatile components, the uranium and thorium accumulate at the beginning of the pegmatitic process. By the end of the process their function in the mineralization is gradually diminishing. If we regard pegmatites according to Fersman, the first is the high temperature type (monazite, orthite) and is characterized by the concentration of thorium in the form of thorite and cerium rare-earth minerals. The second (with rare elements) and the third (boron-fluoride, Li-mucovite) pegmatite types contained most of the uranium in the form of uraninite and yttrium rare-earth minerals, [3.9].

The fifth (sodic-lithium) type has, against a general decrease in the role of the radioactive elements, some increase in their concentrations. It is probably related to the dissolution of higher melting point minerals separated earlier and the redeposition of radioactive elements that have gone into the melt. During the formation of sodic-lithium pegmatites there may be an influx of uranium by metasomatic solutions from outside.

Pegmatites, as distinct from igneous rocks, are characterized by most of the uranium and thorium being concentrated in accessory minerals, and the rock-forming minerals of pegmatites (quartz-feldspathic) often contain less uranium than the parent rock. For instance, Komlev's data indicate that in the microcline of plagioclase pegmatitic veins in Northern Karelia practically all the uranium is concentrated in uranitite, monazite and xenotime. The rock-forming minerals comprise 99.74% of the mass of the pegmatite, but contain only 2.15% of the overall amount of uranium, whereas the accessory minerals comprise 0.25% of the pegmatite mass and account for 98.85% of the uranium.

Typical of pegmatites is the separation of uranium and thorium in rare-earth accessory minerals, where the yttrium and cerium subgroups are markedly split. Moreover, uranium readily enters the lattice of yttrium minerals, and thorium the lattice of cerium minerals as isomorphous impurities.

Uranium pegmatites contain a greater variety of accessory minerals, which distinguishes them from uranium-free pegmatites. The rock-forming minerals are the same as those in the igneous parent rocks and uranium-free pegmatites.

The radioactive minerals of the pegmatites are uranite, thorianite, pitchblende, titanotantaloniobates, titanates of rare-earth elements and thorium, the silicates of titanium, thorium, zirconium, and the rare-earths, the rare-earth phosphates, and uranium-organic compounds.

A feature of these pegmatites is the change in colour and the external features of the minerals exposed to radiation. Quartz acquires a smoky colour; muscovite, which is associated with pitchblende, usually gets darker and becomes brownish. It is recrystallized into fine-scale mica ('decomposed mica'). Bedded together with radioactive minerals, beryl takes on a golden shade, and fluorite becomes black or red-violet. In feldspars surrounding radioactive minerals we can observe zones of reddening and radially diverging jointing.

As a rule, uranium-containing pegmatites are not interesting as a raw material. Uranium is only recovered from them only as a by-product of the extraction of Li, Ce, Be, Ta, Nb, and mica.

The uranium concentration in uranium pegmatites is 0.012%-0.002%. Only some pegmatites that are rich in uranium (0.1% concentration) are economically interesting, while individual deposits may have up to 2% of U_3O_8 (Rex, North-West territories) [3.11].

However, most observations indicate that in the pegmatitic rock crystallized from a melt the original content of uranium was not large. The development of uranium mineralization is mainly related to the post-magmatic processes and depends on the geological conditions of the pegmatite bedding. Fissures in pegmatitic bodies promote post-magmatic processes and lead to uranium mineralization in the form of zones of metasomatic replacement and cavities of solidified filling.

Uranium pegmatites can be found in the formations of all the tectonomagmatic epochs of the Earth's geological history. They are mostly represented on Precambrian platforms (the pegmatites of the Canadian, Baltic, Ukrainian, Aldan shields, and in China, India, Australia, East and South Africa) and more than half of the uranium pegmatitic fields are concentrated there. Paleozoic and Mesozoic shields contain less pegmatite deposits.

3.5 The Geochemistry of Uranium and Thorium in Metamorphism

Metamorphism can be defined as the sum of processes acting below the weathering zone and bringing about the recrystallization of the material in the earth's crust. Metamorphism takes place in hard rocks as a result of changes in the temperature, pressure and chemical conditions. Originally sedimentary and magmatic rocks with different radioactivities vary greatly as a result of the diverse thermodynamic conditions in regional and contact metamorphism, ultrametamorphism, which lead to substantial changes in the forms of uranium and thorium and to their migration (Fig. 3.8).

Two basic factors determine the concentrations of uranium and thorium in metamorphic rocks: (1) the concentration of the elements in the original rock, and (2) the facies of metamorphism.

1. The first factor mainly depends on the formational origin of the initial magmatic or sedimentary rocks. For example, protogeosynclinal structural and facies complexes have very low concentrations of radioactive elements in metamorphic rocks: 1-2 ppm of U and 2–10 ppm of Th (according to Titov, Bilibina, and Kochkin), [3.13]. Clark concentrations of uranium and thorium are typical of protorogenic structural and facies complexes.

The concentration of radioactive elements in the metamorphic mass of various geological epochs is affected most by the original composition of the sedimentary and volcanogenic rocks. Quartzites and limestones have low contents of uranium and thorium. On the other hand, carbonaceous, coaly-like and clay schists and other rocks enriched with organic matter and phosphorus have larger concentrations of dispersed uranium.

2. The progressive metamorphism of rocks provide conditions for migration of uranium and thorium from thermodynamic zones with high temperatures and pressures to those where the physical and chemical parameters are lower (Ermolayev, 1983), [3.7]. In addition, the concentrational dispersion diminishes as the process continues.

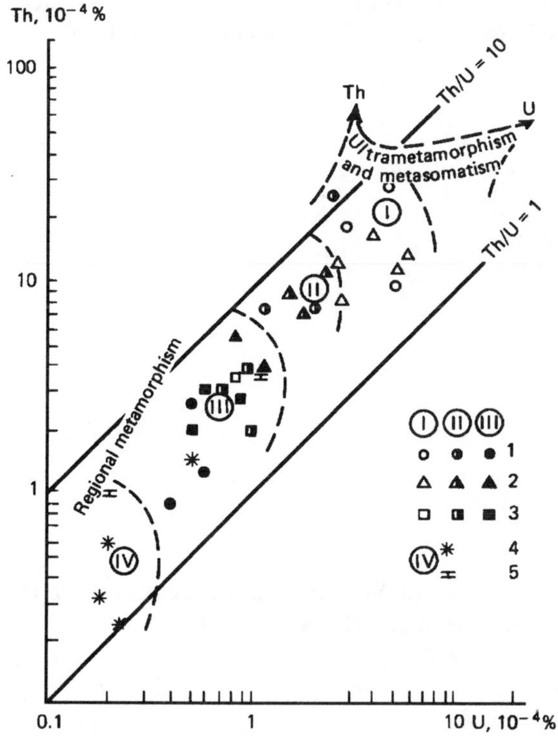

Fig. 3.8. Uranium and thorium metamorphism. Facies of metamorphism:

I—eclogitic; II—granulitic; III—amphibolitic; IV—green-schist. Types of rocks: *1*—gneisses; *2*—crystalline schists; *3*—amphibolites with a different degree of metamorphism (*a*—green-schist, *b*—amphibolitic, *c*—granulitic varieties) (according to Smyslov, 1983)

Where there is low and moderate regional metamorphic grade, the average concentration of uranium is smaller than in the original rocks. However, there is no noticeable dispersion of thorium in this case, in contrast to facies in deep metamorphism. A good illustration can be found from the data of Table 3.5 and Table 3.6.

Local dynamothermal metamorphism, irrespective of the composition of original sediments, is characterized by the same features as the facies of regional metamorphism. In the first place, the concentrations of uranium becomes smaller, while deeper, the same is true of thorium. Secondly, the dispersion of concentrations is lower too.

Table 3.5 Uranium and Thorium in the Metamorphic Rocks of Langéy District of Northern Norway (according to Heier, Adams, 1965), [3.12]

Facies of metamorphism	U, ppm	Th, ppm	Th/U
Epidote-amphibolite	3.45	26.48	7.5
Amphibolite	1.22	9.39	7.0
Granulitic: weakly metamorphosed rocks	0.88	4.09	4.6
Granulitic: strongly metamorphosed rocks	0.39	0.93	2.2

Table 3.6 Uranium (ppm) in Schists and Paragneisses of the Ukrainian Shield (according to Belevtsev), [3.7]

Sediments	Facies of metamorphism		
	greenchist	amphibolite	granulitic
Chemogenic	0.8	0.6	0.4
Terrigenous	2.4	1.7	0.7

In conditions of progressive *contact* metamorphism (low pressure facies) near granites breaking through chemogenic and terrigenous deposits we can observe the migration of uranium. In cases of contact metamorphism, thorium does not tend to be removed. It is partially redistributed among the exocontact zones, which leads to an increase in the concentration heterogeneity.

Regressive Metamorphism

Data on radioactive elements in regressive metamorphism are extremely scarce. Those produced by Titov et al. indicate the existence of uranium and some thorium, in all the facies of regressive metamorphism (Aldan, Baltic and Ukrainian shields). The greatest quantity of uranium and thorium is found in greenschist facies, while the least is found in granulitic facies.

The concentrations of radioactive elements in rocks of the same type that have undergone a similar degree of progressive or regressive metamorphism have much in common. As a rule they are somewhat smaller in the products of regressive metamorphism.

The behaviour of uranium and thorium in zones of contact and dislocation metamorphism in the progressive and regressive stages is similar.

Ultrametamorphism

Ultrametamorphism takes place in zones of the earth's crust and is accompanied by the melting of anatectic material. If the overall composition of the original rocks during the progressive metamorphism is practically unaltered, the system becomes absolutely open for ultrametamorphism. It is actually in this regressive stage that alkalines, silicon, and water begin to enter the rocks that have been somewhat 'desiccated' in the preceding stages of progressive metamorphism. If a physicochemical system is closed, a drop in temperature determines a change in the activity coefficients of the compounds and consequently enhances the activity of the volatile and light petrogenic components.

During the initial stages of ultrametamorphism the uranium and thorium removal continues. As the ultrametamorphism proceeds up to rheomorphism of granito-gneisses and broad palingenesis, the removal of radioelement continues into progressive metamorphism, but is eventually replaced by the opposite, viz. the mobilization of the dispersed uranium and thorium and the formation of melts with enhanced radioactivity. Belevtsev has presented data for the Ukraine, and the average content of uranium in granites that have developed among amphibolite facies is 3.4 ppm, while that in the isofacies of granite in granulitic facies is 1.4 ppm.

Thus, metamorphism leads primarily to a regrouping of the material of which the rock consists including the uranium and thorium dispersed in them. Secondly, most of the water and carbon dioxide, which determine the medium for the transport of the radioactive elements, are removed from rocks that are subject to the progressive metamorphism. Ermolayev (1983) argued that recrystallization and replacement take place at the interface of the rock-forming mineral and the film solution surrounding it. An analysis of the distribution of uranium by fragmental radiography within the bulk of mineral grains indicates that radioactive impurities can be removed from the rock during surface phenomena (desorption into the aqueous carbonic phase formed during metamorphism) or during the recrystallization of the mineral carrier.

During recrystallization the crystallizing substance is purified from impurities, including uranium and thorium.

As the external parameters of metamorphism (pressure and temperature) increase the principal mechanism determining the redistribution of impurities changes from desorption into the separating or filtering aqueous carbonic phase to purification by Ostwald recrystallization, and, finally, to the dissolution (replacement) of the mineral carrier. These assumptions have been verified by experiments on the extraction of uranium from metamorphic rocks by aqueous carbonic solutions. The quantity of desorbed uranium regularly decreases from the products of metamorphism in conditions of green stage to those of granulitic stage. As the temperature and pressure increase further, the metamorphic system separates both uranium and thorium, which can be seen in the decrease of the thorium-uranium ratio from facies to facies. Thorium, which does not have a large-sized oxidized form, as is the case with uranium, is incapable of isomorphism of $(UO_2)^{2+}$ type, and is not on the whole involved in the surface processes which regulate the redistribution of uranium during lower stages of metamorphism. In the later stages, however, the thorium, like the uranium, is removed by Ostwald recrystallization.

During ultrametamorphism the radioactivity increases from the earlier stages to the concluding ones. In the final products, pegmatites of ultrametamorphic complexes, there is a considerable concentration of uranium and thorium together with a corresponding impoverishment in the surrounding rocks. The migration forms of the radionuclides in the melt do not lend themselves to analysis. No ore deposits of uranium and thorium with a metamorphic genesis have yet been discovered. The processes of metamorphism 'expose', as it were, rocks, transmitting the uranium and thorium dispersed in them into forms that can migrate further geochemically.

3.6 The Geochemistry of Uranium and Thorium During Hydrothermal-Metasomatic Ore Formation

The metasomatic and hydrothermal processes that take place with aqueous solutions, are the only endogenic processes in which the concentration of radioactive elements, and primarily uranium, reach industrial concentration. Regional metamorphism creates vast areas in the upper structural layers of the earth's crust with a relatively large content of uranium and thorium and thus prepare the ground for the further concentration. During hydrothermal ore formation, uranium and thorium are separated.

3.6.1 Behaviour of Uranium and Thorium During Formation of Hydrothermal Veined Deposits

Hydrothermal deposits have the richest concentrations of uranium ores, which can reach several tens of per cent (an industrial concentration of uranium is of the order of 0.1%). Over 95% of uranium reserves with endogenic origins are concentrated in hydrothermal deposits. Thus, they are one of the most important sources of uranium [3.14, 3.15].

Hydrothermal deposits of uranium are known for all the metallogenic epochs. They are usually formed at the last stage of hydrothermal process and are subordinated to particular tectonic and magmatic conditions. The most propitious for the formation of uranium ores are the hard blocks of the earth's crust with a relatively stable uplift, in which acid and subalkaline magmatism finds its expression. Characteristic of the ore fields are regional disjunctive disturbances, which have no mineralization in themselves, but serve to conduct and control ores. Ore bodies correspond to the smaller fissures of the rupture and extension, thus edging the large fractures.

As the temperature of the hydrothermal solution decreases, uranium separates from its magmatogenic associates Th, Zr, Ta, Nb. Hence, the uranium titanates, viz. brannerite and davidite, which also contain Th and lanthinides, are found mainly in hypothermal deposits. For the mesothermal and epithermal deposits the ore minerals of uranium are oxides (pitchblende, uraninite) and silicates (coffinite, nenadkevite), and occasionally thucholite.

The numerous classifications now in use classify hydrothermal uranium deposits into three basic types:

(1) plutonogenic, or those that are associated with intrusive complexes of rocks, and are the deepest;

(2) volcanogenic that are associated with the volcanic complexes, varieties which differ in depth;

(3) amagmatogenic or stratiformed deposits; they are not related to any magmatic complexes.

The mineralogical composition of uranium ores and their paragenetic and mineral associations are usually determined by the character of ore-bearing solutions and the influence of the enclosing rocks. There are a fairly large number of mineralogical subtypes of uranium deposits, e.g. sulfide pitchblende, carbonaceous pitchblende, fluoritic pitchblende, arsenic pitchblende, quartz-pitchblende, hydromicaleous pitchblende, and uranobituminous pitchblende. Other subtypes have less practical significance.

The most essential questions for the geochemistry of veined uranium deposits concern the source of the ore body, and the forms of the transportation and factors governing ore deposition. It has been experimentally confirmed and acknowledged everywhere that uranium exists in the hydrothermal solutions in the form of uranyl-carbonaceous complex compounds with potassium and sodium:

$$Me_4\left[UO_2(CO_3)_3\right] \text{ or } Me_2\left[UO_2(CO_3)_2(H_2O)_2\right]$$

The pH of the medium is close to a neutral one. The average temperature of hydrothermal solutions that give rise to pitchblende veined deposits reaches 150 °C on the average, and on the whole does not rise above 200 °C. The temperature pertinent to the formation of hydrothermal and metasomatic ores with uraninite is higher, being 250°-300 °C. The pressure of ore-bearing solutions experimentally determined from gas-liquid inclusions varies from 1000 to 2000 bars, which clearly exceeds the lithostatic loading corresponding to the depth at which the deposits are formed. On the one hand, this indicates the absence of any obvious relationship between the areas of ore formation and the daytime surface, and on the other hand, this shows the influence of deeper zones creating a high excess pressure and maintaining the movement of hydrothermal solutions in the complex hydrothermal system [3.4].

The deposition of uranium minerals is accompanied by the destruction of complex ions and depends on a number of factors, such as a decrease in the redox potential, the change in the pH of the medium, the decomposition of carbonate complexes due to binding CO_3^{2-} into a solid phase as a result of the formation of carbonates or the partial removal of CO_2 during degasation, temperature rise, etc. Thus, the lithological composition of the enclosing rocks is important in the sedimentation of uranium from hydrothermal solutions.

In many cases veins are found in those places which intersect contacts with rocks that change physicochemical conditions of the medium. These rocks may include amphibolites, chlorite or chlorite-epidote schists, diabases, and pyritized rocks (Danchev, Lapinskaya, 1980). The bivalent iron present in these rocks may well be the reducing agent for the uranium. More important for the reduction of the uranium is the presence of organic compounds. The U(VI) dissolved in the carbonaceous medium is then reduced to U(IV), which is hydrolyzed and precipitated upon the corresponding reduction barrier.

The source of orebody is often highly debatable. Geochemical data indicate that the original sources of uranium should be situated along the path of carbonate solutions.

During metamorphism and granitization huge masses of CO_2 are released. Because of low solubility in a granitic melt, they create around an intrusive centre vast zones of hot porous solutions with a high partial pressure of CO_2, which can mobilize ore elements from the enclosing rocks including uranium. In order to extract uranium from rocks to concentrate the ore, a number of conditions regulating the physicochemical properties of the hydrothermal fluids must be fulfilled [3.4].

(1) They should contain fairly large concentrations of carbonic acid.

(2) Their temperature should not be too high as this would lead to the dissociation of the carbonic acid, and, consequently, to a reduction in its chemical activity.

(3) They should have an optimal concentration of alkaline components for the given temperature, i.e. too low pH would reduce the activity of the carbonate anions.

(4) The original uranium content in the enclosing rocks and the forms in which it is present are important. Uranium can occur in acid magmatic rocks that have recrystallized, and in metamorphic rocks or clayey terrigenous sediments containing an adequate amount of mobile forms of uranium.

The distribution of ore components in the enclosing rocks at the 'five-metal' formation shows that the Co, Ni, and Bi come from the enclosing rocks of the basic composition directly at the deposit, by passing from the dispersed state to minerals of hydrothermal veins. At the same time basic rocks cannot be the source of uranium as there is too little of it.

At the same time, the basic rocks cannot be the source of uranium due to its low content. This allows us to speak of the mobilization of uranium by hydrothermal solutions in deeper zones of the earth's crust [3.4].

Ore solutions can penetrate both along open fissures, and along fractured parts of the rocks that are genetically connected with the shattered zones. In these cases the solutions move extremely slowly and come into contact with the enclosing rocks, thus causing metasomatic replacement of the indicated rock. Such deposits are called hydrothermal-metasomatic.

3.6.2 Behaviour of Uranium During Metasomatic Processes

Metasomatic processes can also be regarded as a stage in the mobilization and concentration of uranium. When they affect rocks earlier enriched with uranium, the metasomatosis can result in the formation of industrial deposits. The uranium moves in metasomatizing solutions in the form of carbonate complexes. Therefore, the migration of uranium will be promoted by a weakly alkaline or neutral medium, in which carbonate complexes are most stable. In acid media they will be destroyed because of the poor dissociation of carbonic acid, and in strongly alkaline ones due to the hydrolysis of uranium.

The behaviour of uranium during the formation of metasomatic facies is illustrated by microscopic research using the track method (Omelyanenko, 1983), [3.7].

In skarns the content of uranium is, as a rule, not high or is the same as in the original rock. The forms in which uranium exists depend on the presence of minerals uranium can enter as an isomorphic impurity. When there are high concentrations of rare earths and thorium, skarn deposits with uranium, thorium, and rare-earth elements can appear where alkaline rocks are in contact with limestone, where the ore minerals are represented by pyrochlore, perovskite, betafite, orthite, etc. When the rare-earth elements, zirconium and thorium, with which uranium isomorphism is possible, are at low, the concentrations of uranium in skarns are low too, and practically all of it is found in fissures and intergranular space. This makes it possible to assume that skarn-forming solutions contained fairly high concentrations of uranium.

The behaviour of uranium during albitization is practically the same. Rare-earth albites containing thorium and uranium are known. Here the uranium is also accumulated during the crystallization of minerals with increased isomorphous contents towards it (rare-earth, titanotantaloniobates, cyrtolite, britholite, etc.).

By the end of the albitization process the uranium concentration increases. Part of the uranium is retained in microfissures filled with hematite and chlorite, which reflect the activity of residual (fixed) solutions and indicate relatively high uranium concentrations in albitizing solutions.

Over time albitization is gradually replaced by greisenization. Most researchers mention increased concentration of uranium in greisens. It is mainly deposited in the latest stages of greisen process, between grains and in microfissures. High concentrations of uranium, several hundredths of a percent, have been recorded in gas-liquid inclusions [3.7].

This illustrates the enrichment of greisenizing solutions with uranium and predominance of transport process over deposition.

Thus, alkaline metasomatosis is not essentially an ore-forming process, though it does result in the mobilization of uranium from a fairly large bulk of rock. However, the clogging of porous penetrating zones as a result of the formation of sodic metasomatites or at the expense of tectonic shifts, leads to either carbonaceous-sodic or carbonaceous metasomatosis thus increasing the pressure of carbon dioxide.

This process is not initially uranium-bearing since with the increase in the partial pressure of carbon dioxide the solubility of uranium carbonate complexes rises. As the CO_2 goes to form carbonates, solutions become supersaturated by uranium. The precipitation of uranium leads to the formation of uranium-bearing carbonate metasomatites, spatially connected with albitites.

Metasomatic uranium deposits are widely known. They are spread in the zones of troughs of the Precambrian platforms, where they are coincide with the areas where fields of different age, curves of the axes, sharp limbs, etc. intersect, i.e. where there are the tectonic conditions that promote the slow penetration of ore-bearing solutions, thus increasing their time of interaction with the enclosing rocks.

The conditions for the formation of metasomatic deposits of uranium are: (1) a high concentration of uranium in the solution and the enclosing rocks; (2) a slow crystallization; (3) the presence of the oxidized fluid with a high content of volatile components; (4) intense heat flow.

Deposits of this type are characterized by a complex structure, where together with the typical metasomatic deposits there are metamorphogenic uranium-bearing masses which precede them in time and the younger veins with uranium mineralization which indicate the completion of ore process.

3.6.3 The Behaviour of Thorium in a Hydrothermal and Metasomatic Process

The high-temperature postmagmatic processes can promote the mobilization and concentration of thorium. In these processes thorium is usually associated with rare-earth elements (the lanthanides). Most postmagmatic thorium-containing deposits are connected with pneumatic-hydrothermal or high-temperature hydrothermal processes [3.16].

At present we know of thorium-rare-earth deposits of postmagmatic genesis. As a rule, they are formed from alkaline or subalkaline rocks [3.17], viz: (1) ultrabasic alkaline rocks associated with carbonatite deposits; (2) alkaline and nepheline syenites, which are characterized by the thorium and rare-earth mineralization with increased concentrations of uranium and niobium; (3) subalkaline and alkaline granitoids, where the thorium and rare-earth mineralization is accompanied by Be, Sn, Mo, and W mineralization.

These rock formations have different compositions of lanthanides. In the minerals of the first two formations rare-earth elements of the cerium group predominate, whilst the granitoids have both cerium and yttrium group lanthanides.

Most high-temperature thorium and rare-earth deposits (of pneumato-hydrothermal genesis) were formed as a result of the metasomatic replacement of enclosing rocks. They are characterized by early albitization, K-spathization, aegirinization, amphibolytization, and the formation of skarns. Ore minerals are niobates, tantalo- and titano-niobates (euxenite, pyrochlore, priorite, fergusonite, and others), and the silicates of the rare earths, thorium, and zirconium.

High-temperature hydrothermal deposits are characterized by metasomatic replacements, accompanied by amphibolytization, aegirinization, biotitization, greisenization, and filled open cavities. Niobates, titano- and tantalo-niobates are represented here in small quantities. Rare-earth phosphates (monazite and xenotime) dominate with some rare-earth silicates (orthite, cerite, britholite, miserite, etc.).

The medium- and low-temperature deposits are usually veined with a characteristic change in enclosing rocks (feldspathization, carbonatization, and silicification). Ore minerals are represented by hematite, carbonates and fluorocarbonates of the rare earths (bastnaesite, parisite, synchysite, etc.), and the sulfides of Fe, Cu, Pb, and Zn.

Thus, all hydrothermal-metasomatic thorium- and rare-earth deposits are characterized by a multicomponent ores. The main associates of thorium are the lanthanides. High-temperature metasomatic deposits also have the other important associate of thorium, namely, uranium. Uranium is not, usually, an ore component in hydrothermal deposits proper. In the high-temperature deposits the associates of Th are Ta, Nb, Ti, and occasionally Mo, W, Sn, Be, in medium-temperature, Pb, Zn, Cu. Only one true mineral form of Th occurs in hydrothermal associations, i.e. the silicate. It is either thorite or colloidal complex formations termed ferrithorite. The non-ore elements are usually the alkali metals, primarily K, with some Na. The anion components such as Cl, CO_2, S, and F are important and determine the mobility of Th in postmagmatic fluids.

In post-intrusive supracritical acid gas and porous fluids Th is believed to migrate as fluorine complexes of the type

$$[ThF]^{3+},\ [ThF_2]^{2+},\ [ThF_3]^{+}\ \text{or}\ [Th(F, Cl)_n]^{(2n-4)}$$

(Turovsky and Abakirov, 1979), [3.13].

In postmagmatic mineralization, thorium migrates in alkaline solutions, possibly in the form of complex carbonate compounds with formulae $Th(CO_3)_n^{-(2n-4)}$. Research has revealed that these complexes readily transform to a gaseous form and remain stable as such.

3.7 The Geochemistry of Uranium and Thorium in Exogenic Processes

An important feature of the exogenesis zone is the oxidizing medium. In oxidizing conditions U(IV) is changed to U(VI), thus increasing its mobility. The difference between the chemical properties of U(VI) and Th(IV) causes the division of uranium from thorium in exogenic processes.

3.7.1 The Behaviour of Uranium in Weathering

Weathering is a complex set of processes that disintegrates the rock and chemically transforms them. The behaviour of uranium during weathering, as is the case with other elements, is determined by the following basic factors: (1) the chemical properties of the element; (2) the composition of the rocks and the forms in which the uranium exists in them; (3) the climatic conditions, upon which the physicochemical conditions of the weathering medium depend. They determine the relationship between the aqueous and mineral or fragmental forms of migration.

(1) The interaction between the rock and water where there is active water exchange and good aeration leads to the oxidation of the uranium on the surface of the rock fragments and mineral grains, and further dissolution of uranium (leaching).

(2) The composition of a rock determines its rate of destruction and the release of uranium fixed in the mineral forms. Dispersed uranium which exists in a sorbed state, called 'mobile' uranium most easily converts into solution. The rates of weathering and dissolution of the minerals are different. Dissolution is easiest for the relatively pure oxides of uranium (pitchblende, blacks, and uraninite). Impurities of thorium or titanium (uranothorite, brannerite, davidite, etc.) make the minerals stable to weathering. The solubility of isomorphously bound uranium depends on the stability of the host mineral. For example, uranium isomorphously bound in zircons is practically insoluble in water and migrates with fragmental material.

(3) The climatic conditions determine the rate and character of weathering, as well as the composition of the transporting medium. In regions with an arid climate the mechanical disintegration of the rock predominates while in the north ice cracking is predominant. The rate of chemical weathering increases from north to south.

In a hot arid climate, where the active water exchange has a high carbonate content and alkalinity, uranium is leached rapidly. Water with $pH \geq 8$ are active solvents of the secondary and, to a lesser extent, the primary minerals of uranium. In the oxidation zone of sulphide deposits, sulphuric acid forms and it dissolves any uranium mineral. In the neutral sulphate and chloride solutions, the solubility of uranium minerals is close to the solubility in distilled water.

In a humid climate with an abundance of moisture and vegetation, the water is not very mineralized and mainly contains hydrogen carbonates with nearly neutral pH values. This water does not leach uranium well. The true uranium minerals that can dissolve include the uranyl sulphates, carbonates, and sulphocarbonates. Even in the oxidation zones of sulphide deposits, the excess moisture and intense dilution of the sulphate solutions reduces their solving power. The leaching activity of surface water is improved by dissolved organic acids. Only in humid tropical regions is uranium leached by acid water enriched with dissolved organic matter.

3.7.2 Uranium in Water

The concentration of uranium in surface water often depends on the climatic zone. In rivers and lakes in humid zones with a great amount of precipitation and low evaporation it varies from $n \cdot 10^{-8}$ to $n \cdot 10^{-6}$ g/l. In rivers in arid zones with a low amount of precipitation and rapid evaporation it varies from $n \cdot 10^{-6}$ to $n \cdot 10^{-5}$ g/l, and in lakes it can even reach 1×10^{-4} g/l; this is usually parallel to general mineralization. In large rivers that flow through different climatic zones, the concentration of uranium can change from zone to zone. The Dnieper and the Don have uranium concentrations of the order of $n \cdot 10^{-7}$ in the upper parts and $n \cdot 10^{-6}$ in the lower parts, which flow far to the south. The sources of the Syr-Darya and the Amu-Darya are glaciers and have uranium concentrations of only $n \cdot 10^{-7}$ g/l, which in the middle reaches $n \cdot 10^{-6}$ g/l, and in the lower parts, situated in the dry steppes, it reaches 1×10^{-5} g/l. In the first example we are concerned with horizontal climatic zonality and with its vertical counterpart in the second case [3.18].

The concentration of uranium in oceanic water is fairly stable and is 3×10^{-6} g/l on the average. Deviations only occur due to dilution by river runoff (in the Gulf of Finland it is 0.8×10^{-6} g/l), or in enclosed basins due to evaporation (in the Aral Sea it is $n \cdot 10^{-5}$ g/l) [3.18].

The form in which the uranium exists in water is determined by its composition. In oceanic water with pH from 7.8 to 8.2 and molar concentrations of uranium of 1.43×10^{-8} mol/l, of HCO_3^-, of 2.3×10^{-3} mol/l, and of CO_3^{2-}, of 2.2×10^{-4} mol/l, the most probable form of uranium is the complex ion $[UO(CO_3)_3]^{4-}$ [3.19]. Carbonate complexes of uranium are its main form in river and lacustrine water in arid zones. Dissolved uranium is transported by surface water in the median and north latitudes of the humid zone as simple or polynuclear hydroxyl complexes or complex compounds with fulvic acids or dissolved humic acids. Uranium organic compounds are mainly found in humid tropical areas. Carbonate complexes in these conditions do not exist due to the low solubility of CO_3^{2-} at pH = 5-7.

3.7.3 The Behaviour of Uranium During Sedimentation

The sedimentation of uranium can take place during transportation, or in the final basin of the runoff. The uranium and other elements migrate in coarse rock fragments, as a mechanical suspension, colloid, true solution, or in a living organism. The latter is not significant in the migration of uranium. The relationship between the dissolved, colloidal and suspended forms of uranium during transportation is fairly complicated and depends on many factors. It has been estimated that the concentrations in natural water are most adequately maintained by the corresponding soluble complexes. However, surface water is never free of suspensions or colloids, upon which uranium in solution can adsorb. Indeed, experimental research shows that the suspensions in river water accounts for more than 90% of uranium transported by river.

Continental Sedimentation

The sedimentation of uranium during transportation may occur where the flow slows or where the mechanical suspension is relaxed. Floodplains, swamps and lakes have both mechanical sedimentation and high concentrations of dissolved uranium. Decaying vegetation is important and so the silty sediments of floodplains and lakes, which are rich with organic matter, and low peat bogs have higher uranium concentrations than sediments which do not contain organic substances. In an arid climate such sediments have higher uranium concentrations due to the relatively high content of uranium in the water and due to rapid evaporation.

The presence of organic matter in itself is not a prerequisite for a high concentration of uranium. High peat bogs in a washing regime have lower concentrations of uranium than the enclosing clay sediments; this is a well-established fact. The accumulation of uranium by the organic matter requires a reducing medium, which emerges as organic matter oxides without free access of air (stagnating water basins). The mechanism underlying the binding of uranium by some organic compounds is fairly complicated. Originally the uranium is fixed at the expense of adsorption, which is not strong enough. By reacting with the functional groups of the organic compounds (carboxyl, carbonyl, etc.), the uranyl ion can react to form complex or salt-like compounds. However, the large dimensions of the uranyl ion hinder the emergence of strong organometallic associations. A stable fixation of uranium by an organic compound will only be found in reducing conditions when U(VI) changes to U(IV). The organic matter has a two-fold role in the accumulation of uranium. On the one hand, the absence of free oxygen spent on the oxidation of the organic matter creates a situation to reduce the uranium. On the other hand, the organic matter itself forms stable associations with the U(IV). Thus, organic compounds, particularly vegetative in origin, are very important in the geochemistry of uranium.

Uranium will form weaker bonds when adsorbed by clays and the hydroxides of Fe, Mn, etc. At evaporating interfaces uranium will coprecipitate with carbonates, sulphates, phosphates, vanadates, and arsenates of various metals to form secondary minerals containing the uranyl ion.

Marine Sedimentation

The main source of uranium and other elements in sea water is river runoff and marine abrasion. The dissolved and suspended uranium is mainly deposited on the continental shelf, where practically all the fragmented material is bedded. However, here constant washing and roiling of clastic sediments occur, which under adequate aeration promotes the oxidation and leaching of uranium. Uranium thus dissolved maintains the constant concentration of uranium in sea water.

Only when fragmental material contains accessory minerals stable to weathering (zircon, monazite) do we find uranium-bearing littoral-sea placers. But in these placers uranium is not the basic component and is recovered together with Zr, Th, etc.

A fundamental factor governing the sedimentation of uranium on the shelf is the sorption by organic matter, calcium phosphate (phosphorite), the hydroxides of Fe, Mn and Al, and by clay minerals. The sorption of uranium is at a maximum for pH = 5-6.

The carbonate medium of oceanic water promotes the desorption of uranium, which results in the preservation of sorption equilibrium. Therefore, uranium is not accumulated in larger concentrations than the abundance ratio in oceanic sediments. Thus, sediment formation does not lead to any substantial concentration of uranium. It is only redistributed; it is mobilized from the vast areas of the continents and is moved to the terminal runoff [3.20].

3.7.4 Behaviour of Uranium in the Diagenesis of Marine Sediments

The sediments that have accumulated as a result of sediment formation undergo substantial chemical changes and are affected by micro-organisms. These transformations are called diagenetic.

According to Strakhov [3.21], the earliest diagenetic process is the biogenic absorption of oxygen in silty waters, which changes an oxidizing medium into a reducing one, that is very different from benthic conditions.

Thus, the reduction of sulphates, manganese, iron, vanadium, and uranium begins. As a result, the concentration of uranium in silty water falls. The large difference between the concentrations in silty and bottom waters, frequently as high as two orders of magnitude, promotes a diffusion of uranium into silty waters and its further concentration in the sediment. Under these circumstances the most important concentrating agent of the uranium is the organic matter.

However, as is the case with continental conditions, the presence of organic substances cannot be considered as a marker of uranium concentration. The accumulation of uranium is also affected by the values of Eh and pH, the hydrodynamics, and the rate of sedimentation. The reducing situation is decisive.

Usually, the uraniferous sedimentary rocks, as distinct from non-uranium ones, are characterized by both the abundance of organic matter and by indicators of reducing conditions, viz., the presence of iron sulphides, and the lack of benthic fauna.

Another important uranium concentration process is the formation of phosphorites. However, according to Baturin [3.20], the greatest enrichment with uranium is found in sea phosphorites, which are formed from the bony remains of fish and clay minerals enriched with organic matter and iron sulphides. In other words, the main conditions for uranium-bearing phosphorites are a high concentration of P in the water, the presence of elements that are precipitating agents (Ca, rare-earth elements), and the reducing medium.

Thus, high (in terms of the abundance ratio) concentrations of uranium in marine sediments are formed on the shelf only at the diagenesis stage. Typical examples of these formations are black marine schists.

However, to obtain rich uranium ore deposits another concentration stage is required. This is the process of epigenesis.

3.7.5 The Behaviour of Uranium During Epigenesis

Most exogenous uranium deposits that are important for industrial purposes emerged during epigenesis, or a process superimposed upon sedimentary formations that had appeared earlier.

The emergence of uranium ores in the process of epigenesis requires the combined action of the following principal factors at the same place: (1) a high concentration of uranium in underground water; (2) a favourable hydrodynamic situation; (3) the presence of a reducing interface.

(1) In underground water of the active water exchange as well as in those on the surface, the concentration of uranium is affected by the climatic zone. The concentration grows from 10^{-7}-10^{-6} g/l in humid climates to 10^{-6}-10^{-5} g/l in arid climates. The highest concentration of uranium, up to 10^{-2} g/l can be found where the uranium ores are oxidized, round which aqueous dispersion aureoles form. Where water exchange is hampered, reducing

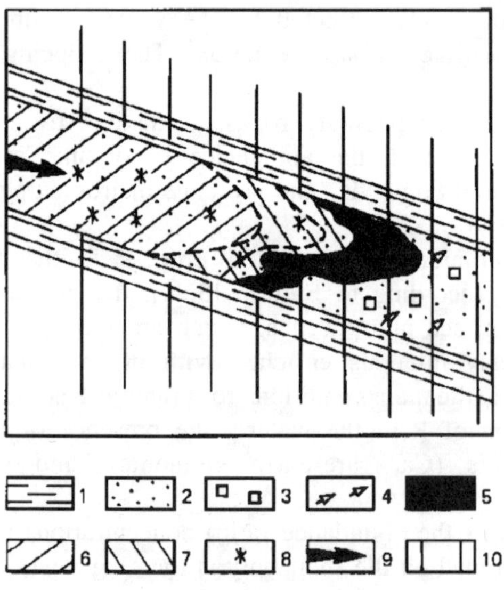

Fig. 3.9. Roll formation chart:

1—Argillaceous-aleurolitic rocks; *2*—sandstones; *3*—pyrite; *4*—vegetational remains; *5*—uranium mineralization; *6*—zone of complete stratified oxidation; *7*—zone of partial stratified oxidation; *8*—hydroxides; *9*—water movement direction; *10*—unaltered grey rocks (according to Kashirtseva, 1970)

conditions cause the transition of U(VI) to U(IV) and its precipitation. Thus, infiltrational deposits of uranium are favoured in waters of active water exchange zone in an arid climate, which wash depleted uranium ores or rocks with a high clark content of uranium (granites, acid effusives, and sandstones). (2) The favourable hydrodynamic situation is caused by the presence of a steep water-permeable bed between two water-confining strata, where the velocity of filtration is commensurable with the time of ore formation.

This can be exemplified by the deposits in grey-coloured terrigenous formations. Here the ore bodies are often associated with paleochannels and are bedded in the gently sloping strata of sandstones that are underlaid and overlapped by clay shales. The oxygen-containing water causes the formation of mottled zones of laminated oxidation. As the quantity of oxygen diminishes, these zones become wedged, and crescent-shaped ore bodies (ore rolls) are formed at reducing interfaces associated with organic matter and iron sulphides.

Ore rolls can move along the strata, thus increasing mineralization (Fig. 3.9).

3.7.6 The Geochemistry of Thorium in Exogenic Processes

Thorium is one of the elements whose dilute aqueous solutions are unstable as a result of hydrolysis. Therefore, during weathering most of the thorium migrates with fragmentary material and suspensions. Thorium forms minerals that are resistant to chemical weathering, which under favourable conditions results in the formation of placers of monazite, thorite, thorianite, columbite, and other thorium-bearing minerals. Known placers include eluvial and eluvial-diluvial types and are associated with eroded laterite and kaolin crusts from the Mesocenozoic age upon granites, syenites, pegmatites, migmatites, and carbonatites. The richest are alluvial and littoral-marine placers in tropical and subtropical climates, which is favourable for the formation of laterite weathering crusts. This is where the rivers wash out ancient eroded crusts and transport the terrigenous material to the coastal zone, where it is rewashed and sorted by specific weight. The lighter minerals are swept out by the tidal currents, while the heavier ones remain, thus enriching beach sands and coastal offshore bars. The length of the path traversed by minerals is different and depends on their weather-resisting properties. Monazite, for instance, can be carried to a distance of 150 km [3.15].

Thus, during erosion, thorium in the form of its own minerals or as isomorphous accessory mineral, is released and migrates as a mineral fragment. The destruction of rock-forming minerals, with which the dispersed thorium is associated, leads to the transition of thorium into the water interacting with the rock. Unstable thorium compounds readily hydrolyze when the pH changes. Hydrolysis products are sorbed onto Fe, Mn, Al hydroxides, on colloidal particles of clay minerals and organic matter that cover the walls of water-conducting fissures in hard rocks, or the surface of mineral grains in friable formations. The newly formed thorium hydroxides are easily leached by carbonaceous solutions and characterize the relatively mobile forms of thorium. Experiments show that during treatment of acidic igneous rocks by the alkalies about 20-30% of thorium contained in the rock can be leached. When the main portion of thorium is contained as accessory minerals, the amount of mobile thorium can drop down to a fraction of one per cent.

Thorium migrates in river runoff mainly with fine suspensions and fragmental material. The concentration of thorium in river water is at least 100 times less than that in colloidal and suspended material. The concentration of thorium in river water is some $n{\cdot}10^{-10}$-$n{\cdot}10^{-8}$ g/l. In underground water it is sometimes higher and reaches $n{\cdot}10^{-6}$ g/l, while in acid sulphate water in the oxidation zone of thorium-bearing rare-earth deposits it is $n{\cdot}10^{-5}$ [3.22, 3.23, 3.24]. Calculations indicate that Th may be present in river waters as simple or polynuclear hydroxy complexes of the type $Th(OH)^{3+}$, $Th(OH)^{2+}$, $Th(OH)_4^0$. In sulphate water we can find the complexes: $Th(SO_4)^{2+}$, $Th(SO_4)_2^0$, $Th(SO_4)_3^{2-}$, etc. Dilution of this water by meteoric one results in a decrease in the concentration of the sulphate-ions, the increase in pH, and the hydrolysis of thorium-bearing complexes. It is also possible to discover thorium in the form of complex compounds with low-molecular fulvic acids (the humid zone) or carbonate complexes (the arid zone). These complexes are presumably rather unstable and readily hydrolyze. Thus, the migration of thorium by water is possible, though, as a rule, it takes place over moderate distances.

The concentration of thorium in the ocean is $n{\cdot}10^{-9}$ g/l. Hence, we may assume that it exists in the form of carbonate complexes. The continuous effect of water enriched with thorium on a natural sorbents can lead to the emergence of epigenetic deposits of thorium, which are, on the whole, rather sparse. The deposits are usually formed where there is endogenic thorium mineralization. For instance, red sandstones are known to exist in cement where together with the secondary hydroxides of Fe(III), 0.001-0.1% of Th can be found. Moreover, some of the thorium does not have any mineral of its own and is dispersed in the minerals of iron, and partly represented by fine-dispersed ferrithorite and thorogummite. The 'iron caps' above thorium-bearing pegmatites and alkaline rocks are often enriched with thorium, which is characterized by primary (residual) and secondary (sorption) mineral forms.

3.7.7 The Distribution of Uranium and Thorium in the Sedimentary Rocks

The geochemistry of uranium and thorium in the exogenesis zone determine the type of distribution of these elements in the sedimentary rocks (Table 3.7).

The variation in the concentrations of the radioactive elements in sedimentary rocks depends on their facies, which reflects the conditions in which they formed.

For example, terrigenous sedimentary formations have a relatively high (close to the abundance ratio) concentrations of uranium and thorium which increase in the order of: conglomerates < gritstones < sandstones < aleurolites < argillites. The association of uranium and thorium in a fine dispersed fraction depends on the role of the clay minerals as sorbents. It is also possible to have a certain concentration of fine grains of accessory minerals, which lead to enrichment, primarily by thorium (Fig. 3.10). In the distribution of uranium in terrigenous sediments organic matter is very important. The ratio Th/U in terrigenous sedimentary rocks is on the average close to the abundance ratio for the earth's crust (3.6-3.7), though in some cases becomes as high as 10.0.

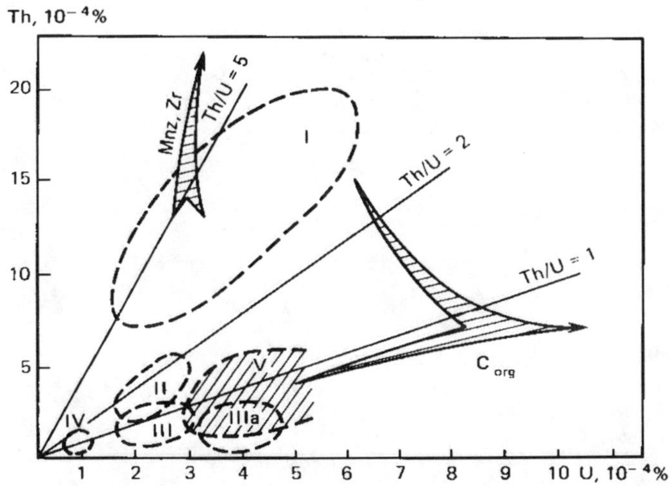

Fig. 3.10. Distribution and accumulation trends of uranium and thorium in sedimentary deposits of the earth's crust:

I—terrigenous; II—siliceous; III—carbonaceous; IIIa—dolomites; IV—saliferous; V—caustobioliths; Mnz—monazite; Zr—zircon (according to Smyslov, 1983)

Table 3.7 Average Content of Uranium and Thorium in Sedimentary Rocks of the Continental Crust (According to Smyslov, 1974)

Groups and types of rocks	U, ppm	Th, ppm	Th/U
	Terrigenous (sandy-argillaceous)		
Conglomerates gritstones	2.4	9.0	3.7
Sandstones, aleurolites	2.9	10.4	3.6
Argillites, clay shales, clays	4.0	11.5	2.4
Quartz conglomerates	6.3	31.0	5.0
Carbonaceous-argillaceous schists	>10-20	15	<1
	Siliceous		
Siliceous shales, quartzites	1.7	2.2	1.2
Clay and siliceous shales	2.8	6.2	2.2
Carbonaceous-siliceous shales	>20-50	10	0.5
	Carbonaceous		
Limestones	1.6	1.8	1.1
Marls	2.8	2.5	0.9
Dolomites	3.7	2.8	0.8
Bituminous limestones	7.8	11.9	1.5
	Saliferous		
Anhydrites	1.0	1.0	1.0
Rock salt	0.9	1.0	1.1
	Caustobioliths		
Hard coal	3.4	4.8	1.4
Peat	2.0-5.0	5.2	1.8
Combustible schists	up to 100-200	10-15	0.5

Table 3.8 Content of Uranium and Thorium in Hard Coal (according to Smyslov, 1974)

Region, basin	Number of analyses	U, ppm	Th, ppm	Th/U
Donetsk	11	1.8	2.8	1.5
Lvov-Volynsk	12	3.0/1.2-6.7	2.5/1.2-8.0	0.8
Ekibastus	6	3.7/2.6-6.1	4.1/2.1-8.5	1.1
Tungus	9	2.4/1.2-5.0	3.7/1.2-6.8	1.5
Tuva	9	2.5/1.6-6.6	2.6/1.2-8.5	1.0
Primorye	20	4.7/2.8-10.8	7.4/3.7-15.7	1.8
Kuzbass	9	4.2/1.4-7.2	8.9/2.8-20.0	2.1
Pechora	7	2.7/1.2-7.9	5.5/1.3-11.5	2.0
Near Moscow	10	6.8/4.6-13.7	8.1/3.7-19.6	1.2
Irkutsk	7	3.5	4.4	1.3
South Ural	6	2.9	4.4	1.8
Kizel	4	2.2	1.9	0.8

Note: average content/fluctuation limit

In marine sedimentary formations, where the main portion of organic matter is accumulated within the confines of the shelf, large concentrations of uranium usually correspond to littoral-sea facies. Carbonaceous rocks are characterized by comparatively low concentrations of radioactive elements. In addition to this, thorium is associated with its terrigenous component (clay), and the uranium may well be brought together with the organic matter, a process known as bituminosity. A minimum concentration of uranium and thorium is typical of halogen formations. Besides, the level of concentrations is also determined by the degree of clayiness.

In caustobioliths the distribution of elements is extremely heterogeneous even within the range of a single basin. The average concentration is usually below the abundance ratio (Table 3.8). At the same time, there are uranium-bearing formations of carboniferous-argillaceous and carboniferous-siliceous schists, where the concentration of uranium is 10-100 times greater than the abundance ratio. Larger concentrations of uranium, sometimes 10-100 times the abundance ratio, are typical of phosphorite formations.

Deposits of Uranium and Thorium in the Sedimentary Rocks

Diagenetic, epigenetic, and polystaged deposits of uranium and thorium in sedimentary rocks exist, which have been formed under the influence of various processes and over a fairly long period of time.

The deposits of uranium in fragmented rocks are widespread on all the continents, cover vast territories, and these reserves are the basis of atomic industry. For all such deposits, the uranium mineralization is associated with zones of bedded oxidation of permeable rocks, which are usually situated at the borders of red (oxidized) and grey (reduced) rocks in the regions with rich organic matter and iron sulphides. Uranium minerals include blacks, hydropitchblende, and occasionally coffinite. A classic example is the Colorado plateau, where 88% of the uranium in the USA is concentrated.

Uranium-bituminous deposits have been found within the oil and gas bearing fields and are associated with the hard bitumens in zones of ancient water-oil interfaces. Sometimes uranium deposits are discovered in red rocks above deposits of oil and are associated with grey-coloured zone reduced by gas from the deposits.

Uranium deposits in carboniferous formations are, as a rule, congruous with the Mesozoic or Tertiary brown coals or lignites in the peripheral zones of the basins close to the alimentation source of underground water. This is where uranium is associated with organic matter and is often accompanied by large concentration of Ge, Be, As, V, Mo, etc.

Uranium deposits in carbonaceous rocks are rather rare. Such ore beds are usually extended along coast lines and correspond to depressions in the floor of shallow-water paleobasins.

Uranium phosphate deposits may be either bedded and nodular in their primary occurrence, or are washed phosphorite pebbles. The content of uranium in phosphorite is not usually more than 0.09% and correlates with P_2O_5.

Placer deposits of uranium and thorium are the basic source of thorium. Monazite placers with littoral-sea genesis are the most industrially important. As a rule, they are polyminerals, and besides monazites, also contain zircon, ilmenite, rutile, columbite, sphene, and garnet. The largest deposits are in India and Brazil. The placer deposits in Brazil stretch for 1600 km along the Atlantic coast and are estimated to involve 150 thousand tonnes, containing 5-6% ThO_2 in monazite. The Indian deposits are along the coast of the Indian Ocean and have about 2 million tonnes of monazite with 8.5% ThO_2 and U_3O_8 from 0.2% to 0.46% [3.17].

Polygenic deposits of uranium and thorium were originally sedimentary rocks enriched with uranium where latest metamorphic and hydrothermal processes redistributed and concentrated radioactive elements. Ancient

(Proterozoic) metamorphosed conglomerates are very important. Another uranium deposit is in carboniferous and flinty shales that have undergone moderate metamorphism.

3.8 The Geochemistry of Uranium and Thorium in the Biosphere

The biosphere is the habitat of living organisms and there is intense interaction between it and all the Earth's geospheres, viz. hydrosphere, atmosphere, and upper lithosphere. The biosphere includes living organisms, the abiogenic component of the habitat (rocks, water, atmospheric gas), and the biogenetic component of the habitat (soils, silts), whose formation is influenced by the activity of organisms. Hence, the geochemistry of elements in the biosphere results from either the direct participation of living organisms, or the geochemical characteristic features of the habitat which appeared due to living organisms. The behaviour of uranium and thorium in the biosphere has been thoroughly studied for soils (habitat) and plants (living organisms). This is the topic of the following section.

3.8.1 Uranium and Thorium in the Soil

'Soil is the upper horizon of the lithosphere involved in the biological cycle with plants, animals, and microorganisms, and is the domain for the highest geochemical energy of all that is alive' (Perelman, 1979), [3.25].

Uranium and thorium behave in soils much as they do in the exogenous zone. The average concentration of radioactive elements in a soil profile depends on the radioactivity of soil-forming rocks. In regions with a dissected topography of substantial importance is the deluvial leaching which alters the original concentration of elements within the confines of even a single soil section. Average contents of uranium and thorium in the soils of this country as derived from regional gamma-spectrometer surveys are given in Table 3.9, [3.26]. At the same time the spread of values is quite large even for the soil-forming rock with the following clark content of radioelements: uranium (0.1-4.5) ppm and thorium (0.1-16.0) ppm. The smallest contents of thorium (from 0.1 to 4.0 ppm) and uranium (from 0.1 to 1.0 ppm) are found in the northern regions of the East European plain (up to 58° n.l.), i.e. in areas with widespread podzol soils. In terms of landscape-climatic zonation they are the regions of tundra and taiga-forest zone where the amount of precipitation

Table 3.9 Average Contents of Uranium, Thorium, and Potassium in Soils of the USSR (according to Boltneva et al., 1980)

Region	U, ppm	Th, ppm	K, %	Th/U	Th/K, 10^{-4}	U/K, 10^{-4}
East Europen and West-Siberian lowlands	1.5	6.5	1.2	4.4	5.7	1.3
Eastern Siberia The Far East	1.4	6.4	1.1	4.6	5.8	1.3
Central Asia Central Kazakhstan	2.3	6.7	1.6	2.9	4.3	1.5

exceeds evaporation, and the leaching of elements from the soil is greater than their accumulation. In the zone where sandy, sandy-loam and turf-podzolic soils are widespread (East European plain) the concentration of uranium and thorium is somewhat higher: (1.0-1.5) ppm U and (4.8-8.0) ppm Th. The southern regions (below 50° south in the West of the USSR and 54° south in the east) have the highest concentrations of radioactive elements: (1.5-3.0) ppm U and (8.0-13.0) ppm Th. This is the black soil and chestnut soils region.

The average clark contents for this country are: U = 1.5 ppm; Th = 6.5 ppm, and Th/U = 4.4.

Within each soil profile, elements are redistributed between the genetic horizons [3.27]. If the average concentrations of uranium and thorium and their ratio along the whole profile of the soil are functions of their concentrations in the soil-forming rocks, then they are spread along the individual genetic horizons depending on the character and intensity of soil formation, which grows from north to south (for the Northern Hemisphere there is a marked differentiation of the elements along a profile in the same direction). In all cases the behaviour of uranium and thorium is associated with two basic processes: the accumulation in the mineral fraction inherited from the soil-forming rocks and the accumulation from soil waters. The relationship between these processes is largely determined by the form in which the element exists in the soil-forming rocks. Most uranium isomorphously enters into accessory minerals that are stable to chemical weathering, which determines its low concentrations in soil waters and insignificant role of sorptive forms. The zone of biogenic accumulation is important for the highest horizons of soil.

The accumulation of uranium usually becomes apparent in the upper humus and gley horizons and is associated with the presence of organic matter and reducing media. The latter factor is decisive. This is a good illustration of the dependence of the distribution of radionuclides on the landscape. In soils that have developed in well drained terrain on watersheds, we find that the uranium is even leached out of humus horizons. Most of it correlates with the mineral and fragmental fraction. In soils in the same region but in depressions with stagnant conditions uranium accumulates from soil waters. The soils in the northern and middle lattitudes with humid climates are characterized by similar regularities in the migration of uranium, the only difference being its intensity due to the rate of the soil-formation processes. In forest-steppe and steppe zones the accumulation of uranium also depends on factors such as evaporation, due to which carbonaceous horizons enriched with uranium form in the soils, and sometimes true minerals of U(VI) may be released. This behavior of uranium in soil should be taken into account when looking for secondary oreoles of dispersed uranium.

Most of the thorium in soils is associated with the mineral fraction. Within a solid profile we find thorium leached from the upper horizons (Group A) and a larger concentration in the illuvial horizon (B_2). The thorium mainly moves with silty fraction. Frequently the upper humus horizons have low Th/U ratios due to accumulation of uranium on an organic matter and the removal of thorium. However, we mentioned earlier that thorium can migrate by water within the biosphere. Hence, the distribution of thorium within the soil profile also depends on the form in which it exists in the soil-forming rock. Thorium migrates slower where in the soil forming rocks it is mainly bound in rare-earth minerals resistant to chemical weathering. The thorium which passes into the soil water is rapidly sorbed by the colloids of Fe, Mn, organic compounds, and clay minerals, and therefore migrates, in the main, in the silty fraction. Where the thorium concentration corresponds to the abundance ratio, the water migration features are well defined. For example, in the podzol soils of the taiga, thorium is concentrated in the A_0 turf horizon and in the horizons that coincide with the water table. Dissolved organic matter is important in the water migration of thorium. Thorium is rapidly leached from carbonaceous soils in arid zones by the soil waters and redeposited on the secondary carbonates within the soil profile.

3.8.2 Distribution of Uranium and Thorium in Plants

The migration of elements in the system consisting of soil and plant is evaluated using the coefficient of biological absorption (CBA), which is defined as the ratio of the concentration of the element in the plant to its concentration in the root layer of the soil.

Table 3.10 Migration Coefficients of ^{238}U and ^{232}Th in the Biosphere [3.24]

Coefficient	System (relationship)	Uranium				Thorium			
		Mountain tundra	Mountain taiga		Forest steppe	Mountain tundra	Mountain taiga		Forest steppe
K_l	$\frac{\text{water}}{\text{rock}}$	0.09	0.07	1.7	0.9	0.02	0.07	0.2	—
K_d	$\frac{\text{sediment}}{\text{water}}$	1.1	5.6	1.6	5.8	2.7	2.2	3.4	—
K_a	$\frac{\text{sediment}}{\text{rock}}$	0.09	0.4	2.7	5.2	0.05	0.02	0.7	0.7
K_i	$\frac{\text{soil}}{\text{rock}}$	0.3	0.2	1.7	5.0	0.03	0,02	1.2	0.5
CBA	$\frac{\text{plant}}{\text{soil}}$	0.21	0.17	0.08	0.005	0.05	0.30	0.03	0.46

The coefficient for uranium is usually less than 1, [3.28]. This means that living plants do not accumulate uranium. The value of the coefficient for uranium depends on the form in which it exists in the soil. Where uranium is mainly in a mineral-fragment forms, the coefficient is higher. Where the most of the uranium is accumulated by humus, the coefficient is lower. This is because the acidic root extractions of plants leach uranium from minerals, but are unable to decompose humates, which are insoluble in acid media.

Uranium accumulation by plants is nearly the same in all climatic zones. There are differences between plants as to the accumulation of uranium. The best accumulators of uranium are the mosses, especially their aqueous forms. As the uranium concentration in the soil increases the CBA decreases, which indicates the existence in the root system of plants of 'biological barriers' to uranium.

Thorium is accumulated as intensely as uranium. The data of Titayeva and Taskaev (1983) [3.28] show that over 70% of plants being analyzed that grow in different climatic zones have a coefficient of biological absorption for ^{232}Th higher than that for ^{238}U. We may also observe distinctions between the accumulation of thorium by plants. In the mountain-tundra terrain the indicating species for thorium is the dwarf-birch. As is the case with uranium, the thorium also shows its 'threshold effect' when it is absorbed by plants. Depending on the composition of the soils and species of the plants, the threshold concentration of thorium in the soil varies from $n\cdot10^{-3}$ g/g to $n\cdot10^{-2}$ g/g.

The peculiarities of the migration of uranium and thorium in biosphere are given in Table 3.10 containing the coefficients of migrations in the following systems: rock-water (K_x), sediment-water (K_d), sediment-rock (K_a), soil-rock (K_e) and the plant-soil coefficient (CBA) for particular regions of mountain tundra, mountain taiga, and forest steppe. In the case of tundra and taiga region No. 1 the main part of uranium and thorium in soil-forming rocks and soils is connected with accessory minerals. For these regions, the element migration by water is minimal, while the CBA values are maximal.

REFERENCES

3.1. Seaborg, G.T., Katz, J.I., 1975. The Chemistry of the Actinide Elements. John Wiley and Sons, New York, London, Sydney, Toronto.

3.2. Vdovenko, V.M., 1969. The Radiochemistry Today. Atomizdat, Moscow.

3.3. Makarov, E.S., 1973. Isomorphism of Atoms in Crystals. Atomizdat, Moscow.

3.4. Naumov, G.B., 1978. The Fundamentals of Physicochemical Model of Uranium Ore Formation. Atomizdat, Moscow.

3.5. Shcherbina, V.V., 1972. The Fundamentals of Geochemistry. Nedra, Moscow.

3.6. Smyslov, A.A., 1974. Uranium and Thorium in the Earth Crust. Nedra, Leningrad.

3.7. Sun, S.S., 1980. Lead Isotopic Study of Young Volcanic Rock from Mid-ocean Ridges, Ocean Islands and Island Arcs.—Phil. Trans. Roy. Soc., London, **A 297**, No 1430, 409-445.

3.8. The Problems of Radiogeology, 1983. (Ed. N.P. Laverov, G.B. Naumov). Nauka, Moscow.

3.9. The Principal Features of Uranium Geochemistry, 1963. (Ed. A.P. Vinogradov). USSR Acad. Sc. Publ., Moscow.

3.10. Titayeva, N.A., Berman, I.B., Ishbulatov, R.A. et al., 1987. The Track Method in Studying the U-distribution in Recent Volcanic Rocks.—The Track Method in Geology and Geography. Vladivostok, 100-111.

3.11. Heier, K.S., Adams, Y.A.S., 1965. Concentration of Radioactive Elements in Deep Crustal Material.—Geochim. et. Cosmochim. Acta, **29**, No 1, 53-61.

3.12. Radioactive Elements in Rocks. (Proc. of 1st All-Union Radiogeochemical Conf., May 1972), 1975. Nauka, Novosibirsk.

3.13. Uranium Hydrothermal Deposits, 1978. (Ed. F.I. Volfson). Nedra, Moscow.

3.14. Danchev, V.I., Lapinskaya, T.A. Deposits of Radioactive Raw Material, 1980. Nedra, Moscow.

3.15. Mineralogy and Geochemistry of Rare and Radioactive Metals, 1987. Energoatomizdat, Moscow.

3.16. Geology of Postmagmatic Thorium and Rare-earth Deposits, 1972. (Ed. V.A. Nevskii). Atomizdat, Moscow.

3.17. The Search Methods of Uranium Deposits, 1969. Nedra, Moscow.
3.18. Evseeva, L.S., Perelman, A.I. and Ivanov, K.E., 1974. The Geochemistry of Uranium in the Zone of Hypergenesis, Atomizdat, Moscow.
3.19. Lal, D., Nijampurkar, V.N., Somayajulu, B.L.K. et al., 1976. Silicon Activities in Coastal Waters of the World Ocean.—Limnology and Oceanography, **21**, 285-293.
3.20. Baturin, G.N., 1975. Uranium in Sea Sedimentation Today. Atomizdat, Moscow.
3.21. Strakhov, N.M., 1962. The Fundamentals of Lithogenesis Theory. Vols. I and II. USSR Acad. Sc. Publ., Moscow.
3.22. Drozdovskaya, A.A., Melnik, Yu.P., 1968. New Experimental and Calculation Data of Thorium Migration in Hypergene Conditions.—Geochemistry, No 4, 402-410.
3.23. Titayeva, N.A., Maslov, V.I., Taskaev, A.I. et al., 1982. Hydrogeochemistry of Thorium Isotopes in the Zone of Active Water Exchange.—Water Resources, No 5, 77-82.
3.24. Dementiev, V.S., Syromyatnikov, N.G., 1965. About the Form of Thorium Isotopes Finding in Ground Waters.—Geochemistry, No 2, 211-218.
3.25. Background Radioactivity of Soils and Rocks on the Territory of the USSR, 1980. (Ed. by L.I. Boltneva, I.M. Nazarov, Sh.D. Fridman).—Trudy Instituta Prikladnoi Geophiziki, **43**, Gidrometeoizdat, Moscow.
3.26. Titayeva, N.A., Taskaev, A.I., 1983. Migration of Heavy Natural Radionuclides in Humid Zone Conditions. Nauka, Leningrad.
3.27. Heavy Natural Radionuclides in the Biosphere. Migration and Biological Action on Population and Biocoenosis, 1990. (Ed. P.M. Aleksakhin). Nauka, Moscow.

4 The Isotope Geochemical Methods Employing Primary Radionuclides and Their Stable Decay Products (Radiogenic Isotopes)

'The introduction of a numerical geological measurement of time is the first step towards a quantitative geology'.

V.I. Vernadsky

4.1 The Basic Methodological Principles

Radiogenic isotopes are the daughter nuclides, or decay products of long-lived parent radioactive nuclides. Geochemistry uses mainly the following pairs of parent and daughter nuclides:

^{238}U-^{206}Pb	^{87}Rb-^{87}Sr
^{235}U-^{207}Pb	^{147}Sm-^{144}Nd
^{232}Th-^{208}Pb	^{167}Re-^{167}Os
^{40}K-^{40}Ar	^{176}Lu-^{176}Hf

On the basis of the law of radioactive decay and formula (1.18) it is possible to write a relationship between the number of atoms of parent nuclide (P) and daughter nuclide (D)

$$D = P(e^{\lambda t} - 1) \tag{4.1}$$

The equation (4.1) is fundamental and is used when radiogenic nuclides are analyzed. Since the number of atoms of parent and daughter nuclides contained in a mineral now can be determined analytically, the relationship between P and D becomes a formula for estimating the age of minerals (t). From (4.1), we get

$$t = \frac{1}{\lambda}\ln\left(1 + \frac{D}{P}\right) \tag{4.2}$$

In order to determine the age of a mineral it is essential to follow some rules:

(1) Throughout its history the mineral must be a closed system as regards D and P; there should have been no removal or addition of daughter or parent atoms.

(2) At the moment of crystallization the mineral should not have any daughter nuclei, i.e. all the D atoms in it were formed as a result of the decay of P atoms, and so are radiogenic. Natural minerals do not, as a rule, obey the first condition. Atoms D within a mineral are admixtures, since they differ, with few exceptions (the Nd-Sm system), by their chemical and crystallochemical properties from P atoms. Besides, D atoms may be shifted from the initial position of P in the mineral lattice as a result of radioactive release if this was induced by alpha decay or spontaneous fission (Ch. 2). A complete loss of atoms leads to a new starting point for the geochronological clock. Thus, the possibility of 'opening' systems in various epochs of the existence of a mineral imposes particularly rigid conditions on the geological material under consideration.

The second condition usually remains unaccomplished. At the time of crystallization, the mineral could always capture some daughter atoms present in the original substrate. We shall designate the number of the atoms that were captured during crystallization as D_0. The expression (4.1) will then be

$$D = D_0 + P\left(e^{\lambda t} - 1\right) \tag{4.3}$$

The stable isotopes of a particular element are analyzed using mass-spectroscopy, which determines the ratio between isotope atoms differing in mass. Therefore, an analysis makes it possible to determine the ratio between the radiogenic isotope D_1 and the nonradiogenic isotope D_2. By dividing both sides of the equation (4.3) by D_2 and replacing D by D_1, we get the final expression used analytically:

$$\frac{D_1}{D_2} = \left(\frac{D_1}{D_2}\right)_0 + \frac{P}{D_2}\left(e^{\lambda t} - 1\right) \tag{4.4}$$

In order for (4.4) to be used it is necessary that we should know the value of $\left(\frac{D_1}{D_2}\right)_0$, i.e. the isotope ratio in the original admixture of the element—'the common D'. The establishment of this value is one of the most important tasks of geochronology, since common analytical methods cannot differentiate between radiogenetic D atoms and those captured during crystallization.

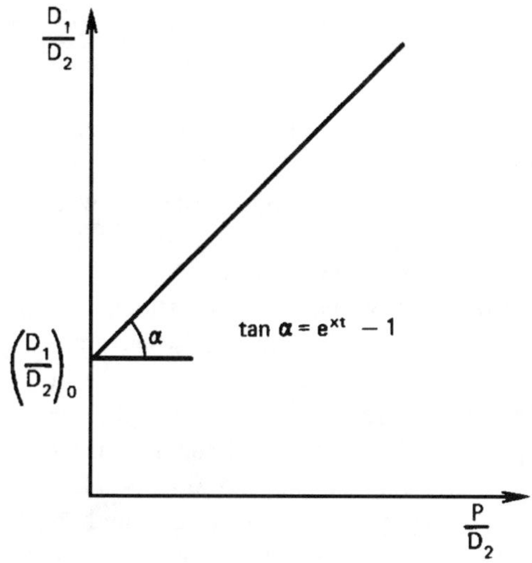

Fig. 4.1. Isochrone

The isotope composition of the 'common' daughter element D was determined earlier from a mineral of the element itself, that is paragenetically associated with the mineral employed in the geochronological analysis. For instance, for U-Pb or Th-Pb systems, the common lead was estimated from halenite. However, the danger that these minerals may crystallize from substrates with different isotope compositions of lead means there is still an error introduced while determining $\left(\frac{D_1}{D_2}\right)_0$.

Nowadays, a more reliable and widely used method is that of *isochrones*. An isochrone is the straight line connecting the points of isotope compositions of minerals of the same age that differ in their contents of the parent element P. Indeed, (4.4) is an equation of a straight line in $\frac{D_1}{D_2}$ and $\frac{P}{D_2}$ coordinates (Fig. 4.1). An isochrone will intersect the abscissa at the point corresponding to the composition of 'the common element D', i.e. $\left(\frac{D_1}{D_2}\right)_0$. The slope of the straight line α will be the function of the age of the mineral, viz. $\tan\alpha = e^{\lambda t} - 1$. Thus, the method of isochrones makes it possible to determine the age of a geological body from several syngenetic specimens and to establish the initial isotopic ratio of the daughter element in the substrate. If the specimens have different ages, the points of their isotopic compositions will not lye on one straight line.

The ratio $\left(\frac{D_1}{D_2}\right)_0$ is interesting, since it yields data on the geochemical history of the source from which the geological body under analysis was formed. These ratios are widely employed in isotopic geochemistry to investigate the genetic properties of igneous rocks. With this aim in view, an isotopic systematization was developed, i.e. the distribution of the isotopic ratio as a function of the genesis of the rock source. More often it becomes feasible to construct a global isotopic systematization, which can be used to determine to which basic reservoirs of the Earth does the source belong to, viz., the continental crust, the oceanic crust, the depleted and non-depleted mantle, as well to discover the contamination of other reservoirs or their mixing [4.1].

The time interval that can be determined by a certain method of radiogenic isotopes is established by the following factors. The lower limit depends on the accuracy in determining the minimal quantity of radiogenic isotope, while the upper limit depends on the possible loss of elements during early metamorphosis and equals the age of the Earth.

4.2 The Potassium-Argon Method

4.2.1 Potassium-Argon Method of Dating

The radiogenic isotope of argon ^{40}Ar is formed as a result of K-capture during the radioactive transformation of ^{40}K. The diagram showing the transformation processes is given in Fig. 4.2 [4.2]. Thus, only 11.2% of ^{40}K atoms transforms into argon, which should be taken into account when estimating the age by the ratio of $^{40}Ar/^{40}K$ (Ch. 1). Formula (4.2) in the case of double decay will be

$$T = \frac{1}{\lambda_K + \lambda_\beta} \ln\left(\frac{\lambda_K + \lambda_\beta}{\lambda_K} \cdot \frac{^{40}Ar}{^{40}K} + 1\right) \tag{4.5}$$

where ^{40}Ar is the number of atoms of radiogenic argon; ^{40}K is the number of atoms of ^{40}K in the sample; $\lambda_\beta = 4.92 \times 10^{-10}$ year^{-1} is the constant of β-decay; $\lambda_K = 5.81 \times 10^{-11}$ year^{-1} is the constant of K-capture. ^{40}Ar is determined by mass spectroscopy. ^{40}K is estimated from the total content of potassium according to

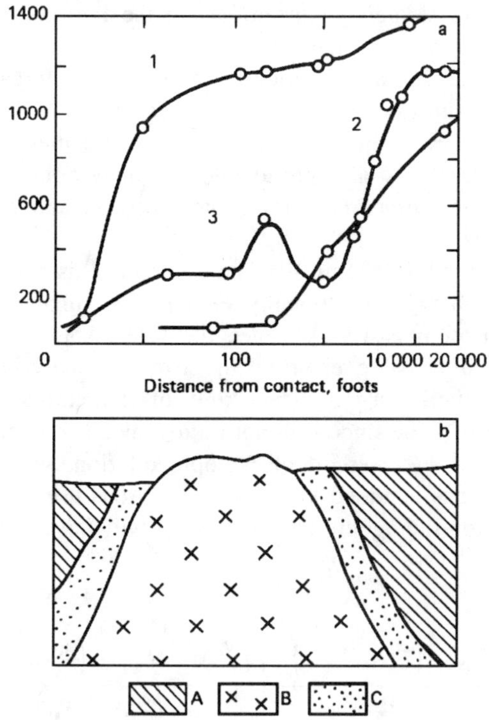

Fig. 4.2. The loss of argon in zone of intruzive contact:
A–late intrusion; B–ancient rocks; C– zone of loss of argon;
1–amphibole; *2*–biotite; *3*–feldspar (Doe, Hart, 1963)

the ratio $^{40}K/K = 1.167 \times 10^{-4}$. The concentration of potassium can be determined by a method, the error of which can be compared with that in determining the argon concentration (flame photometry, atomic absorption, X-ray fluorescence, mass-spectrometric method of isotopic dilution, chemical means, etc.).

In nature the following K and Ar isotopes (in atomic percentage of the total sum of isotopes) have been found:

$$
\begin{array}{ll}
^{39}K = 93.2581 & \left.\begin{array}{r} ^{40}Ar = 99.600 \\ ^{38}Ar = 0.063 \\ ^{36}Ar = 0.337 \\ ^{40}Ar/^{36}Ar = 295.5 \end{array}\right\} \text{atmospheric argon} \\
^{40}K = 0.01167 & \\
^{41}K = 6.7302 &
\end{array}
$$

Potassium is an alkaline element of group 1A in the Periodic Table. It is a petrogenic element and is found in many common minerals, such as feldspars, micas, sylvite, and carnallite. Its concentration in amphiboles may reach a whole percentage. Potassium is substantially lithophilic, accumulating in the upper part of the earth's crust during the evolution of magmatic melts. Potassium is contained in endogenic fluids and participates in metasomatic processes. During halogenesis it forms minerals, for instance, sylvite, and carnallite. It is found in clay minerals.

Argon is a noble gas and is chemically inert. Argon exists in potassium minerals in a free state, being mechanically enclosed in mineral structures. The radius of the argon atom is 1.9 Å. Since the basic isotope of argon in the atmosphere is ^{40}Ar, a possible error during isotopic analysis is due to contamination by atmospheric argon. This kind of contamination can be assessed by the presence of ^{36}Ar, since it is not radiogenic. The overall amount of argon is established by the method of isotopic dilution, when a known amount of one of its isotopes is added as a tracer. If ^{38}Ar is used as the tracer, the calculation formula for radiogenic argon ($^{40}Ar_{rad}$) will have the following expression [4.3]:

$$^{40}Ar_{rad} = \frac{P_{^{38}Ar_{tr}}\left(^{40}Ar/^{38}Ar\right)_s}{P_s}\cdot\left[\frac{\left(^{40}Ar/^{36}Ar\right)_s-\left(^{40}Ar-^{36}Ar\right)_a}{\left(^{40}Ar/^{36}Ar\right)_s}+\frac{\left(^{40}Ar/^{36}Ar\right)_a-\left(^{40}Ar/^{36}Ar\right)_{tr}}{\left(^{40}Ar/^{38}Ar\right)_s\left(^{38}Ar/^{36}Ar\right)_{tr}}\right],\ cm^3/g \tag{4.6}$$

where $^{38}Ar_{tr}$—quantity of tracer, cm^3; P—mass of specimen, g; index is "t","s" and "a" mark of isotope ratios of tracer, specimen and atmosphere, respectively.

It is much more difficult to determine errors due to the preservation of argon and potassium and capturing of argon during crystallization of a mineral, or the supply of potassium as a result of more recent processes. The possibility of the argon capturing is important in the analysis of young samples. The loss of elements depends on the crystal chemistry of the mineral and its history. Argon is mainly lost in the thermal metamorphism. For this reason the age of igneous rocks produced by the K-Ar method will be the age of the latest thermal metamorphism, as a result of which the Ar was entirely or partially lost. The materials used in analysis are the various widespread potassium minerals, e.g. micas, orthoclases, amphiboles, plagioclases, and sometimes pyroxenes. The most widespread are the micas. Depending on the temperature, minerals lose argon in different amounts, which is naturally reflected in the age.

Figure 4.2 shows the loss of argon in different minerals depending on the distance from the intrusive body [4.4].

The isochrone method is appropriate in determining the age where isotopic data available for the minerals with different potassium contents and for the same geological body are plotted in a straight line (Fig. 4.3). The most accurate data are obtained when the age is determined for igneous minerals. The dating of a rock in general is undesirable, since it depends on the size of the grains and the migration of elements on their surface. Volcanic rocks are analyzed from the rock as a whole, if the size of grains is fairly small, or from a glass, sanidine, amphibole, or biotite.

K-Ar dating can be applied to sedimentary rocks in two cases. The time of sedimentation can be established from authigenic glauconites that have been

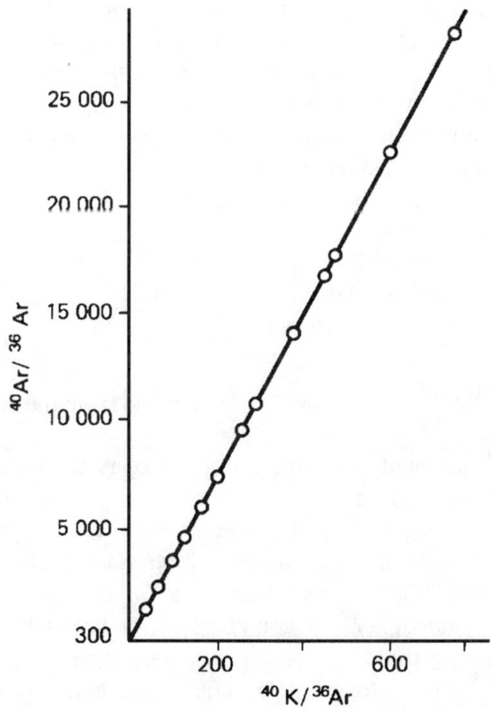

Fig. 4.3. K-Ar isochrone of white micas from the deposits of Nana, South-West Africa, 530 mln years (according to J.C. Hunziker, 1984)

properly preserved. The analysis of minerals of terrigenous rocks (sandstones, conglomerates) helps to determine the age of fragmental minerals, and thus to establish the watershed regions of ancient sedimentation basins by correlating the figures obtained with the age of the surrounding igneous rocks. It is desirable that metamorphic complexes are dated by either K-Ar or Rb-Sr methods.

The former yields the date of the most recent thermal metamorphism, while the latter helps us to establish the date when the rocks were formed. Argon capture is important for young volcanic rocks. This is particularly true for products of underwater eruptions.

4.2.2 $^{40}Ar/^{39}Ar$ Method of Dating

This method is a K-Ar dating variant, in which the ^{39}Ar content is assessed rather than the potassium content, the ^{39}Ar being formed from ^{39}K by neutron activation according to the reaction of ^{39}K (n, p) ^{39}Ar. An advantage of the method is that the analysis requires one sample using a mass-spectrometer and other argon isotopes can be analyzed simultaneously. A sample and a control with a known age are sealed in quartz ampoules and placed in a reactor, where they are irradiated by fast neutrons.

The dating formula can thus be represented as [4.2]:

$$t = \frac{1}{\lambda}\ln\left[1 + J\left({}^{40}\mathrm{Ar}_{\mathrm{rad}}/{}^{39}\mathrm{Ar}_{\mathrm{K}}\right)\right] \tag{4.7}$$

where $J = \frac{^{39}\mathrm{K}}{^{40}\mathrm{K}} \cdot \frac{\lambda_K + \lambda_\beta}{\lambda_e} \Delta T \int \Phi(\varepsilon)\sigma_\varepsilon \, d\varepsilon$; J is calculated from the controls using (4.8), Φ_ε is the flux of fast neutrons with energy ε; σ_ε is the cross section of reaction ^{39}K (n, p) ^{39}Ar with energy ε; ΔT is the duration of irradiation.

Some of the isotopes of argon may also emerge as a result of other nuclear reactions, particularly if the sample contains a high Ca/K ratio viz., ^{40}K (n, p) ^{40}Ar; ^{39}K (n, nd) ^{37}Ar; ^{40}Ca $(n, n\alpha)$ ^{36}Ar; ^{40}Ca (n, α) ^{37}Ar.

The contamination by atmospheric argon is established by using ^{37}Ar, which is formed exclusively during the radiation process. The samples that have been irradiated are used for argon release. A promising method is to release the argon with gradual heating the sample and measuring the argon at each temperature stage (the method of step burning). The last stage is the complete melting of the sample. The age that is determined in the complete melting should correspond to the usual K-Ar age. This method should yield for each

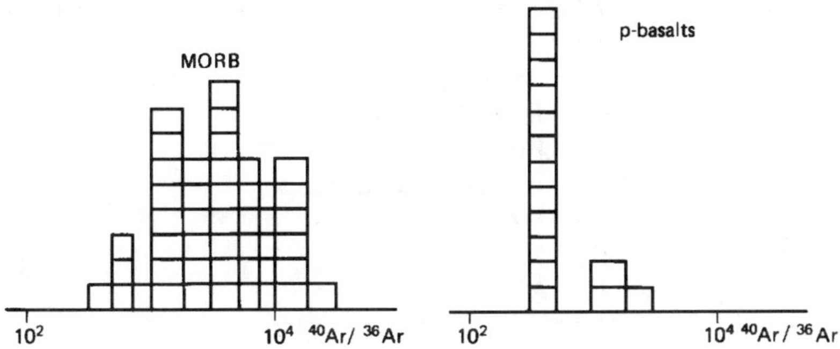

Fig. 4.4. Histogram of $^{40}Ar/^{36}Ar$ distribution in M- and P-basalts (according to A.B. Verkhovsky, et al., 1986)

sample a set of data that, in the form of a histogram, is called the age spectrum. If the sample has always been a closed system, all the data will be the same. This method has been comprehensively applied to meteorites and volcanic rocks, where most of the K is in glassy form. In crystalline rocks the process is complicated by the various forms in which K exists and due to the different threshold values of the ^{39}Ar activation energy. In order for the age to be estimated, an isochrone is constructed from the isotopic composition of the various temperature fractions of a sample [4.5].

4.2.3 Isotope Systematics of Argon and Helium

The ratio of $^{36}Ar/^{40}Ar$ in the main terrestrial reservoirs differ from the meteoritic ratio. Moreover, within each reservoir the ratio varies within certain limits. The isotopic systematic employes the dependence of the isotopic ratio on the affinity of the rock source to one of the main terrestrial reservoirs. When analyzing a rock whose source is unknown, this systematic helps us evaluate the genetic peculiarities of the rock.

In most papers the isotopic systematic of argon is considered together with that of helium (Table 4.1).

The variations in helium and argon isotopes depend in many ways on the same factors. In the ratio $^{40}Ar/^{36}Ar$, ^{36}Ar is primary and ^{40}Ar is mostly radiogenic, having been formed from K. This can be clearly seen from Table 4.1 [4.5] where the maximum value of $^{40}Ar/^{36}Ar$ occurs in the earth's crust, within which most of the earth's potassium is concentrated. In the ratio $^{3}He/^{4}He$, ^{3}He is primary, while ^{4}He consists of two components: primary and

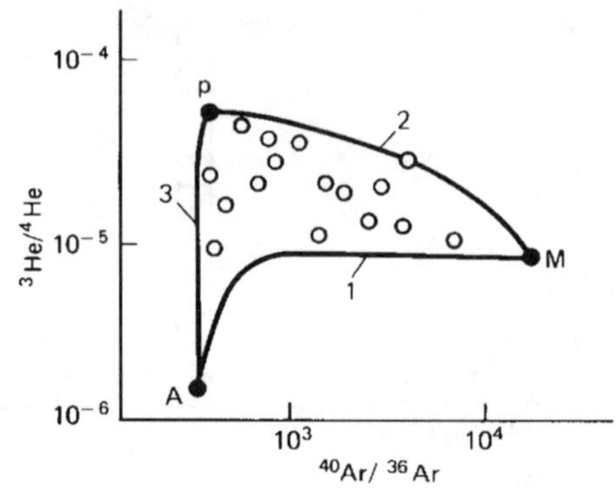

Fig. 4.5. Isotopic composition of He and Ar in pallial reservoirs (M and P) in the Earth's atmosphere (A) (according to A.B. Verkhovsky et al., 1986):
1-3 –lines of mixed components (circle—basalts)

radiogenic helium, the latter having been formed in earth due to alpha radiation from natural radionuclides, mainly the ^{238}U and ^{232}Th decay series. The lithophile elements of uranium, thorium and potassium are characteristic for the earth's crust, and therefore ^{4}He is dominant in this particular reservoir.

As an example of Ar-He isotopic systematic let us consider the data on the distribution of the ratios $^{40}Ar/^{36}Ar$ and $^{3}He/^{4}He$ in the basalts of the ocean today [4.6] (Figures 4.4 and 4.5). According to the present concepts, the basalts of the mid-oceanic ridges (MORB) originate from the uppermost layer of the mantle, which is depleted by the lithophile elements (the 'depleted mantle' or M reservoir).

Given this data it is possible to conclude the following. (1) The greater radiogenic helium concentration at the boundary between the upper mantle and the oceanic crust corresponds to the greater radiogenic argon concentration. A flux of helium passing through the mass of the upper mantle, the oceanic crust and water helps to maintain a fairly homogeneous distribution of its isotopes and the high concentrations in the glasses of MORB, which exceed the concentration in 'hotspot' in the basalts by two orders. The homogeneity of the isotopic composition points to the deepness of the flux, which evidently seeps along the gigantic channels of the mid-oceanic ridges. By analogy with He, the Ar observed in the M basalts also characterizes mantle isotopic composition. The larger variations are due to a certain contamination by atmospheric argon dissolved in sea water, which cannot influence the isotopic composition of the helium. (2) The lower concentration of radiogenic helium in 'hotspot' is evidence of its great depth. At the same time the isotopic

Table 4.1 The Isotopic Composition of Ar and He in Various Reservoirs of the Earth [4.4]

Reservoirs	$^{40}Ar/^{36}Ar$	$^{3}He/^{4}He$
The originally homogeneous Earth of 4.55×10^9 years ago	$\approx 10^{-4}$	2.4×10^{-4}
Non-depleted mantle	400	5×10^{-5}
Depleted mantle	2000	1.1×10^{-5}
The earth's crust	10^7	2×10^{-8}
The atmosphere	195.6	1.39×10^{-6}

composition indicates whether there has been contamination by atmospheric argon, which could only take place during contact with either the atmosphere or sea water. If we suggest that 'hotspot' had been caused by the melting of fragments of the oceanic crust that entered the mantle by subduction (Turkott, 1984), it would not be difficult to explain the contamination by atmospheric argon. Besides, it would be obvious that contamination would not be insignificant since its concentration in the mantle plume is much greater than that in the atmosphere itself.

4.3 The Rubidium-Strontium Method

The rubidium-strontium method is based on the radioactive transformation of ^{87}Rb by β-decay with a half-life of 5.0×10^{10} years, to form ^{87}Sr. The system links two elements which substantially differ in their chemical properties.

Rubidium is an alkali element in group IA of the Periodic Table. It is a chemical and crystallochemical analogue of K ($R_{Rb^+} = 1.48$ Å, $R_{K^+} = 1.33$ Å) and isomorphously replaces it in all minerals. It does not have any minerals of its own. Like K, Rb is a lithophile element. During magmatic differentiation it accumulates in residual melts, and enters endogenic fluids. The main rubidium-bearing minerals are micas and orthoclases. The greatest concentrations of Rb have been found in lithium micas and pegmatites. Variations in the K/Rb ratio are very great indeed (from $n \times 10$ to $n \times 1000$).

Strontium is an alkaline-earth element in Group IIA in the Periodic Table. It is chemically and crystallochemically close to Ca ($R_{Sr^{2+}}$ = 1.13 Å, $R_{Ca^{2+}}$ = 0.99 Å). Strontium has a coordination number of eight in minerals. In the basic rock-forming minerals, strontium isomorphously replaces Ca in plagioclases, and in apatites of the accessory minerals. The replacement of Ca^{2+} by Sr^{2+} in pyroxenes is hampered because Ca has a coordination number of six in pyroxenes. In K-feldspars Sr^{2+} can replace K^+. Meanwhile Si^{4+} is replaced by Al^{3+} in a silicic acid tetrahedron. The replacement of K^+ by Sr^{2+} in micas is complicated because potassium has a coordination number of 12.

In distinction to Rb and K, Sr is removed from magmatic melts during the crystallization of primary and secondary Ca-plagioclases. Hence, the value of the Rb/Sr ratio is increased with differentiation of magmatic melts. According to [4.8] the Rb/Sr ratio is on average 0.23 for the earth's crust. In basalts it falls to 0.06, but in granites it rises to 0.53 on average.

4.3.1 Rubidium-Strontium Method of Dating

In accordance with (4.4) the Rb/Sr ratio is expressed as

$$({}^{87}\text{Sr}/{}^{86}\text{Sr})_t = ({}^{87}\text{Sr}/{}^{86}\text{Sr})_0 + \frac{{}^{87}\text{Rb}}{{}^{86}\text{Sr}}(e^{\lambda t} - 1) \tag{4.8}$$

The age t is thus determined as

$$t = \frac{1}{\lambda} \ln\left(1 + {}^{87}\text{Sr}_{\text{rad}}/{}^{86}\text{Sr}\right) \tag{4.9}$$

where $\lambda = 1.42 \times 10^{-11}$ year^{-1} is the decay constant of ^{87}Rb.

^{86}Sr was chosen as the stable isotope, since its content is close to the ^{87}Sr content.

The isotopic compositions of Rb and Sr in terrestrial matter is [4.4]:

$$^{85}\text{Rb} = 72.15\% \qquad ^{84}\text{Sr} = 0.56\%$$

$$^{87}\text{Rb} = 27.8\% \qquad ^{86}\text{Sr} = 9.86\%$$

$$^{87}\text{Sr} = 7.02\%$$

$$^{88}\text{Sr} = 82.56\%$$

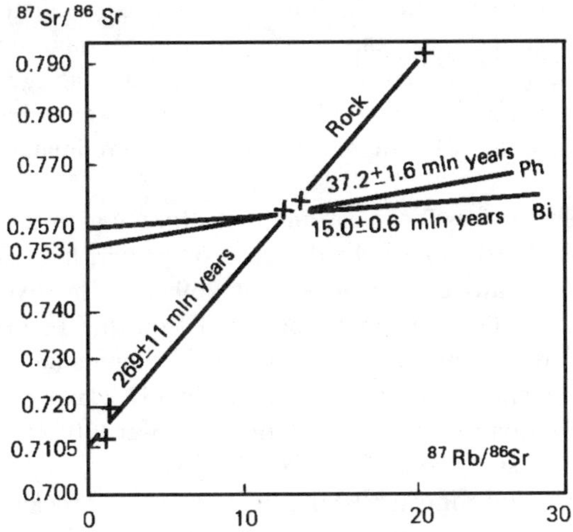

Fig. 4.6. Rb–Sr isochrons for bulk rock samples and micas from the granites of Rotondo, Central Alps (according to E. Yëger, 1980)

The isotopic ratios and the contents of Rb and Sr in samples are usually established by mass-spectrometer in the method of isotopic dilution. Rubidium-strontium dating is somewhat restricted by: (1) The long half-life of ^{87}Rb which leads to a slow accumulation of radiogenic ^{87}Sr. Hence tertiary rocks are the youngest to which the Rb-Sr method is applicable. (2) The high content of strontium in the earth's matter and the high content of ^{87}Sr in it determines a prominent geochemical background for ^{87}Sr/^{86}Sr.

Consequently, the radiogenic additive is hard to distinguish from the background of common strontium, particularly in young samples. The best samples are those with a high Rb/Sr ratio.

Minerals used for rubidium-strontium dating are in descending order of the Rb/Sr ratio [4.1]: biotite, phengite, muscovite, chlorite (which emerged from biotite), and orthoclase, in which the K/Rb ratio is often identical to the rock as a whole. Minerals in which we have a large concentration of Sr, rather than Rb are plagioclase, apatite, fluorite, epidote, and garnet. Consequently, to establish the minimal (Tertiary) age only biotite and phengite should be used.

A particularly significant topic is the behaviour of the Rb-Sr system during metamorphism, which is usually superimposed upon ancient igneous rocks. Research [4.1] has shown that even at relatively low temperatures there is a partial exchange of Sr between minerals. Radiogenic Sr removed from a biotite does not, as a rule, leave the rock, and concentrates in neighbouring minerals containing Ca (apatites, plagioclases). That is why frequently the age determined by Rb-Sr dating for the rock as a whole proves to be more reliable than if applied to separate minerals. The sample should be approximately 30 kg by weight [4.2].

Figure 4.6 illustrates the ages for micas and rocks for the granites of Rotonda and the Central Alps (Yeger, 1984). In the case of a rock in general it was possible to obtain an isochrone with an age of 269 ± 11 mln years and an initial ratio of 0.7105 ± 0.0012. Phengite emerged from the Lepontine phase of Alpine metamorphism. Its threshold crystallization temperature is 500°, and therefore it dates the culmination of metamorphism (37.2 ± 1.6 mln years). The threshold crystallization temperature of biotite is lower (300 °C), and hence the cooling of rock is dated, i.e. 15 ± 0.6 mln years.

Rb-Sr dating is successful for establishing the age of lunar rocks, meteorites, volcanic and plutonic rocks, the epochs of metamorphism. Sedimentary masses have been dated using authigenic clay minerals (illite, smectite). It is impossible to date sedimentary masses by the rock as a whole since they can contain residual strontium in the composition of the fragmental fraction.

4.3.2 Isotope Systematics of Strontium

Research into the initial ratio of $(^{87}Sr/^{86}Sr)_0$ in magmatic and metamorphous rocks in different regions of the Earth has revealed that its distribution depends on the evolution of the sources of rocks in the main terrestrial reservoirs. It is necessary to take into account that the $(^{87}Sr/^{86}Sr)_0$ ratio which reflects the Rb/Sr ratio in the systems that preceded the source, is an integral value. It is hardly feasible to determine the number of these systems, the time of the existence and Rb/Sr ratio in each of them. Even though we are concerned with the result of action of these systems, i.e. with a one-stage model, the strontium isotopic systematic makes it possible to identify the origin of the rocks in a main terrestrial reservoir, and in a number of cases to estimate the contamination by the material of another reservoir or mixing of various sources.

The evolution of the isotopic composition of strontium in magmatic rocks of the Earth is shown in Fig. 4.7 [4.7]. The diagram divides all the points into three zones. The middle narrow band corresponds to the evolution of the isotopic composition of strontium in the primary homogeneous undepleted mantle.The initial value of $^{87}Sr/^{86}Sr$, which had existed in this mantle for 4.55 bln years ago has been established by the meteoritic substance and is estimated

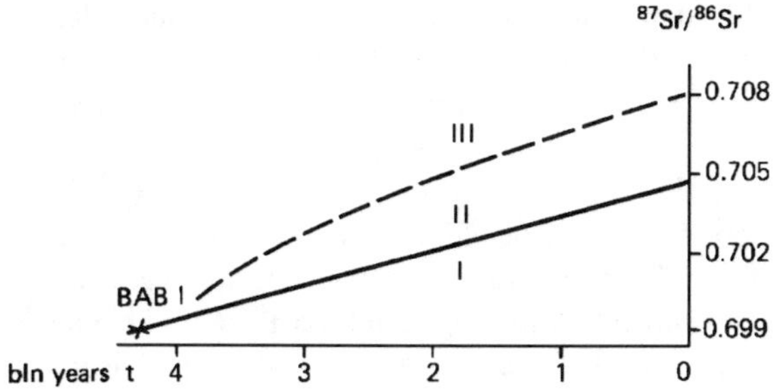

Fig. 4.7. Evolution of isotopic compositions of strontium.
Fields: I–oceanic upper mantle; II–continental upper mantle; III–continental crust (according to Yu.A. Balashov, 1983)

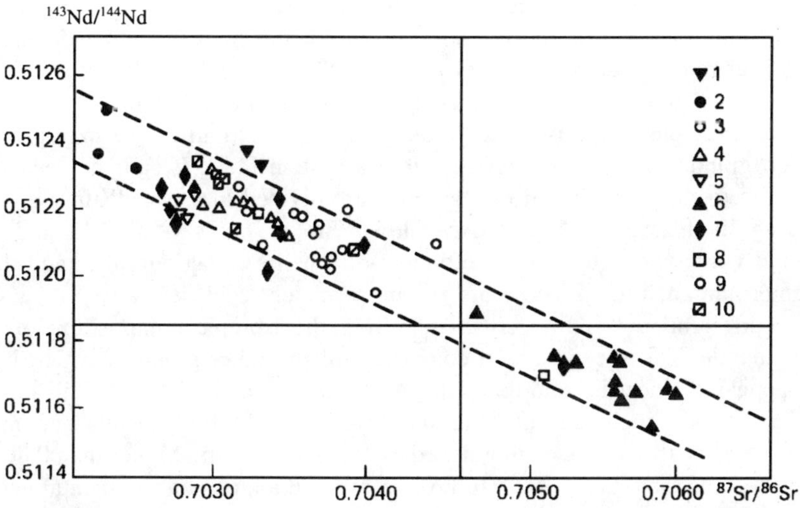

Fig. 4.8. Correlation between the isotopic composition of strontium and niodium for young igneous rocks of the world:
1–mid-Oceanic ridge (Indian Ocean); *2*–mid-Atlantic ridge; *3*–the Hawaii; *4*–Iceland; *5*–Ascension Is; *6*–Kerguelen Is; *7*–Galapagos Is; *8*–Tristan da Cunha; *9*–Bouvet France; *10*–rocks from the wells on the ocean floor; dotted lines restrict the main pallial correlation, solid lines show SHUR values (according to D.De Paolo, 1976)

as 0.69898 (BABI). Proceeding from this value it becomes possible to calculate $^{87}Sr/^{86}Sr$ in the present-day undepleted mantle, UR (Union Reservoir) which is equal to 0.7045, as well as the ratio Rb/Sr = 0.30. A somewhat similar value was obtained from the correlation diagram $^{143}Nd/^{144}Nd$-$^{87}Sr/^{86}Sr$ (Fig. 4.8), [4.8].

The derivatives of the depleted mantle are situated below the UR line. For instance, oceanic tholeiites (mid-oceanic ridges, ocean islands, and ocean floor) have the values of $^{87}Sr/^{86}Sr$ less than 0.7045. The average value for most of the tholeiites of the mid-Atlantic ridge and East-Pacific elevation is 0.7026, while for ferro-basalts of the ocean floor and alkaline basalts of the ocean islands it is 0.7032. The effusives of island arcs and primitive series of developed arcs have low isotopic ratios of strontium (0.7027-0.7044). Other rocks in developed island arcs and continental margins are remarkable for their high ratios of $^{87}Sr/^{86}Sr$, sometimes exceeding 0.7047.

The continental rocks occupy in the diagram of Fig. 4.4 a vast region above the UR line. As distinct from the oceanic igneous rocks, continental igneous rocks have variable isotope compositions, which reflects their complex genesis. The upper part of the diagram contains figurative points that correspond to the derivatives of typically crustal continental magmatism. The points in the proximity of the UR line correspond to mantle derivatives. Only several points for the traps of some of the regions fall within the field of depleted mantle, which illustrates the complicated evolution of their source. $^{87}Sr/^{86}Sr$ of sea water is equal to 0.7090-0.7091 and corresponds to continental values.

As an example of the isotopic systematization of strontium we may present data on granites which were brought into a system [4.7]. It can be seen that according to the $^{87}Sr/^{86}Sr$ ratio all granites are divided into two distinct groups differing in their genesis. Relatively low $^{87}Sr/^{86}Sr$ ratios of 0.7045 and the minimal variations in the ratio are typical of tonalites, plagiogranites and their metamorphic analogues. These are the most ancient acid derivatives of a still poorly differentiated sialitic crust, inheriting the isotopic composition of the upper mantle. The second group comprises all the other granites, which have only appeared since the middle of the Archean period. These have still higher $^{87}Sr/^{86}Sr$ ratios which are substantially differentiated. Bimicaceous and muscovite granites are always distinguished by the isotopic ratio higher than 0.7045. It may well be that from 3.0-2.8 billion years the thickness of continental nuclei was sufficient for melts in the upper part of the crust to reflect its composition. In this connection more ancient material has been treated, which resulted in the observable variations.

4.4 The Samarium-Neodymium Method

4.4.1 Samarium-Neodymium Method of Dating

The basis of the samarium-neodymium method is the radioactive transformation of ^{147}Sm into ^{143}Nd by alpha decay. ^{147}Sm has a half-life of 1.06×10^{13} years, and, hence, the method is applicable to analyze events that occurred long ago. It was successfully used to date lunar rocks, meteorites, Archean volcanites, etc. [4.1, 4.10]. The method cannot be used to establish the age of relatively young rocks since radiogenic neodymium accumulates too slowly.

The samarium-neodymium system links two elements of the same chemical group, the lanthanides, and they have almost the same chemical properties.This makes the samarium-neodymium method advantages since it does not depend on secondary processes that can differentiate the elements in the system. When analyzing metamorphic rocks, the rubidium-strontium method makes it possible to establish the age of metamorphism, while the samarium-neodymium method helps us to ascertain the age of original rocks. The equation for the isochrone is

$$^{143}\mathrm{Nd}/^{144}\mathrm{Nd} = \left(^{143}\mathrm{Nd}/^{144}\mathrm{Nd}\right)_0 + ^{147}\mathrm{Sm}/^{144}\mathrm{Nd}\cdot\left(e^{\lambda t}-1\right) \tag{4.10}$$

Samarium is a trivalent rare-earth element with Ri = 1.04 Å. High concentrations of samarium are found in rare-earth minerals containing mainly light and medium lanthanides (monazite, cerite, bastnäsite, orthite, and others). Sm enriches the melt during differentiation. The average contents of Sm in basalts are 6.9 ppm, in granites 8.5 ppm, and in the sedimentary rocks, 7 ppm.

Neodymium is also a rare-earth element with $R_{Nd^{3+}} = 0.995$ Å. Its concentration also grows during differentiation. The average content of Nd in basalts is 32 ppm (with great variations), in granites it is about 55 ppm. The highest concentrations of neodymium are in rare-earth minerals containing light lanthanides.

4.4.2 Isotope Systematics of Neodymium

The isotopic systematics of neodymium is only possible for geochemical processes leading to the differentiation of Sm and Nd. In spite of the fact that both are rare-earth elements that share many geochemical properties there are mechanisms by which they can be separated. The rare-earth elements are lithophilic and during melting accumulate in the residual melt. However, owing to differences between the distribution coefficients the melt becomes richer in 'lighter' lanthanides with respect to 'heavy' ones. The extraction of light lanthanides from the mantle during melting is proportional to the loss of alkaline elements (K, Rb, Cs). This means that the ratio of light to heavy lanthanides (or medium ones), e.g., La/Yb, La/Sm, and Nd/Sm, will be proportional to Rb/Sr, and consequently to $^{87}Sr/^{86}Sr$ and inversely proportional to $^{143}Nd/^{144}Nd$.

The variation in $^{143}Nd/^{144}Nd$ is insignificant over time due to the long half-life of ^{147}Sm. It has been suggested [4.1, 4.8, 4.9] that instead of the ratio $^{143}Nd/^{144}Nd$ the difference εNd between the isotopic composition of neodymium in the sample and the isotopic composition of the evolving averaged original chondritic reservoir CHUR (Chondritic Uniform Reservoir) be used

$$\varepsilon Nd = \left[\frac{\left(^{143}Nd/^{144}Nd\right)_m}{\left(^{143}Nd/^{144}Nd\right)_{CHUR}} - 1 \right] \times 10\,000$$

The current value of $(^{143}Nd/^{144}Nd)_{CHUR}$ was estimated to correspond to 4.55×10 bln years, given a primary value of $^{143}Nd/^{144}Nd = 0.512638$ $^{147}Sm/^{144}Nd = 0.1967$, which corresponds to Sm/Nd = 0.313, the value obtained from meteoritic assays [4.1].

The isotopic systematics of neodymium is not informative in itself since the variations of $^{143}Nd/^{144}Nd$ in time are very small compared with the error of determining $(^{143}Nd/^{144}Nd)_{CHUR}$, and because of the differentiation of Sm and Nd. More important is the inverse correlation between $^{143}Nd/^{144}Nd$, $^{87}Sr/^{86}Sr$ established in [4.8]. Figure 4.8 shows the principal mantle correlation which brings together the isotopic compositions of young basalts of the ocean floor and ocean islands and reflects the evolution of mantle substance.

This correlation is the basis of the concepts about the evolution of the Earth. In principle the Nd-Sr isotopic correlation may be the result of two opposite processes: the mixing of two homogeneous reservoirs, which differ in the ratios of Rb/Sr and Sm/Nd, or fractionation in an original homogeneous reservoir. The latter assumption is more consistent with the observed facts. The ratio $(^{143}Nd/^{144}Nd)_0$ obtained by dating terrestrial rocks using the samarium-neodymium method lie along the evolution line, which would be impossible if two reservoirs were mixed. The second assumption would imply a relationship with the petrochemical and geochemical data: the melting of an originally homogeneous mantle leads to two markedly different reservoirs: the continental crust richer in Rd than in Sr and poorer in Sm than in Nd, and the depleted upper mantle with the converse ratios of elements (complementary to the continental crust relative to the original mantle).

From the Nd-Sr correlation diagram we obtain $(^{87}Sr/^{86}Sr)_{CHUR}$ = 0.7045 corresponding to $(^{143}Nd/^{144}Nd)_{CHUR}$.

When points for a rock in the Nd-Sr diagram lie outside the principal mantle correlation band, this means that the rocks form due to the melting of sialitic crust, or it is a contamination by sialitic material, sea water, the influence of pallial metasomatosis, etc. As an example of this interpretation of isotopic data, it is possible to adduce the data of Zhuravlev et al. [4.6] concerning the Kuril Islands. The points for contemporary volcanic rocks from the Kuril island arc partly lie within the principal mantle correlation band, but partially deviate to the right. The authors explain this anomaly by the fusion of hydrothermal mantle in the subduction zone, and the transversal zonality, by the presence of the depleted mantle in the rear zone of the mantle wedge, and, presumably, by the influence of the fluid stream.

4.5 The Uranium-Thorium-Lead Method

4.5.1 Methods of Dating

^{238}U, ^{235}U, and ^{232}Th produce a lot of radionuclides which decay into each other, and thus form radioactive decay series or radioactive families. The end decay products for all the three series are lead isotopes. When studying slow processes the relatively short-lived intermediate members of the series can be disregarded and the simplified systems of ^{238}U-^{206}Pb, ^{235}U-^{207}Pb, and ^{232}Th-^{208}Pb are considered. The basic parameters of these systems are

Parent nuclide, P	Content in % of the sum of isotopes	The period of half-life, years	The constant of decay, years	Daughter nuclide, D
^{238}U	99.2739	4.468×10^{9}	1.55125×10^{-10}	^{206}Pb
^{235}U	0.7204	0.7038×10^{9}	9.8485×10^{-10}	^{207}Pb
^{232}Th	100	1.4008×10^{10}	4.9475×10^{-11}	^{208}Pb

The $^{238}U/^{235}U$ ratio is constant and equals 137.88 (with the exception of the Oclo uranium ores, Ch. 7). Hence, instead of an isotopic analysis of uranium we need only establish its overall content. Besides radiogenic isotopes ^{206}Pb, ^{207}Pb and ^{208}Pb, there is another stable isotope, viz. ^{204}Pb. The concentration of ^{204}Pb does not change with time and it is considered to be nonradiogenic. However, it is possible that it is formed as a result of an extremely slow alpha decay of ^{200}Hg with a half-life of 1.4×10^{17} years [4.11].

Lead is in Group IIB of the Periodic Table and is a chalcophile element. Its most widespread mineral is galenite, PbS. However, lead can also manifest its lithophile properties. Owing to its large ionic radius ($R_{Pb^{2+}} = 1.20\ \mathring{A}$), it can replace potassium ($R_{K^{2+}} = 1.33\ \mathring{A}$) in orthoclases and calcium ($R_{Ca^{2+}} = 1.01\ \mathring{A}$) in plagioclases. The first is most probable. Lead cannot enter the lattice of zirconium, or of uranium and thorium minerals due to the great difference between the charges of their ions and ionic radii. The lead in the minerals of these elements unaltered by recent processes is mainly radiogenic.

Using Eq. (4.4), each of the systems U-Pb and Th-Pb can be written as

$$^{206}Pb/^{204}Pb = \left(^{206}Pb/^{204}Pb\right)_0 + ^{238}U/^{204}Pb \times \left(e^{\lambda_{238}t} - 1\right) \tag{4.11}$$

$$^{207}Pb/^{204}Pb = (^{207}Pb/^{204}Pb)_0 + ^{235}U/^{204}Pb \times \left(e^{\lambda_{235}t} - 1\right) \tag{4.12}$$

$$^{208}Pb/^{204}Pb = (^{208}Pb/^{204}Pb)_0 + ^{232}Th/^{204}Pb \times \left(e^{\lambda_{232}t} - 1\right) \tag{4.13}$$

whence we get expressions for determining the age

$$t_{206} = \frac{1}{\lambda_{238}} \times \ln\left[\frac{\left(^{206}Pb/^{204}Pb\right) - \left(^{206}Pb/^{204}Pb\right)_0}{\left(^{238}U/^{204}Pb\right)} + 1\right]; \tag{4.14}$$

$$t_{207} = \frac{1}{\lambda_{235}} \times \ln\left[\frac{\left(^{207}Pb/^{204}Pb\right) - \left(^{207}Pb/^{204}Pb\right)_0}{\left(^{235}U/^{204}Pb\right)} + 1\right]; \tag{4.15}$$

$$t_{208} = \frac{1}{\lambda_{232}} \times \ln\left[\frac{\left(^{208}Pb/^{204}Pb\right) - \left(^{208}Pb/^{204}Pb\right)_0}{\left(^{232}Th/^{204}Pb\right)} + 1\right] \tag{4.16}$$

After dividing Eq. (4.13) by Eq. (4.12) and eliminating the uranium we get

$$\frac{\left({}^{207}\mathrm{Pb}/{}^{204}\mathrm{Pb}\right)-\left({}^{207}\mathrm{Pb}/{}^{204}\mathrm{Pb}\right)_0}{\left({}^{206}\mathrm{Pb}/{}^{204}\mathrm{Pb}\right)-\left({}^{206}\mathrm{Pb}/{}^{204}\mathrm{Pb}\right)_0}=\frac{{}^{235}\mathrm{U}\left(e^{\lambda_{235}t}-1\right)}{{}^{238}\mathrm{U}\left(e^{\lambda_{238}t}-1\right)}$$

or

$$\left({}^{207}\mathrm{Pb}/{}^{206}\mathrm{Pb}\right)_{\mathrm{radiogenic}}=\frac{1}{137.8}\times\left(\frac{e^{\lambda_{235}t}-1}{e^{\lambda_{238}t}-1}\right) \quad (4.17)$$

In order to estimate age by Eq. (4.18) special tables are used. Various uranium- and thorium-bearing minerals are employed for analysis. Since the main requirement for a sample to be used for the evaluation of age is that the system is closed, the U-Th-Pb method uses minerals that are most stable to the action of external processes, e.g. zircons, sphenes, monazites, orthites, and uraninites. At present the most reliable data come from zircons.

In principle, if the system remained closed throughout the existence of the mineral, the ages from the four equations (4.15, 4.16, 4.17, 4.18) should be the same. However, as a rule, even zircon data usually diverge. Both zircons and other uranium minerals have the following succession of discordant ages which is due to either the release of radiogenic lead or the addition of radioactive elements [4.12]:

$$\frac{{}^{207}\mathrm{Pb}}{{}^{206}\mathrm{Pb}}>\frac{{}^{207}\mathrm{Pb}}{{}^{235}\mathrm{U}}>\frac{{}^{206}\mathrm{Pb}}{{}^{238}\mathrm{U}}>\frac{{}^{208}\mathrm{Pb}}{{}^{232}\mathrm{Th}}$$

When there is discordance, we cannot obtain a reliable age for a single mineral, even using the most reliable system ^{207}Pb/^{206}Pb. A series of cogenetic minerals is analysed. The data is interpreted by various graphical methods.

If during the existence of the uranium mineral it has not released or received U and Pb, the ages determined from the ^{207}Pb/^{235}U and ^{206}Pb/^{238}U ratios must coincide. Such ages are called concordant [3.8]. In ^{207}Pb/^{235}U-^{206}Pb/^{238}U coordinates the points of concordant ages will lie along a curve, called the concordia (Fig. 4.9). This line represents the development in time of the ratio ^{207}Pb/^{206}Pb. If the mineral has recently lost lead or there has been an influx of uranium, then the point will lie below the concordia (point *A*). If, however, the reverse occurred, the points lie above the concordia (point *C*), but this happens very seldom. If there has been a single redistribution of elements in

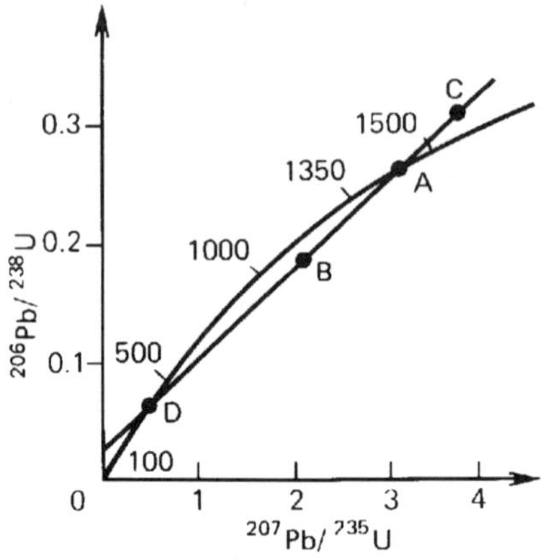

Fig 4.9. Concordance and discordance (see A, B, C, D values in the text)

a coeval U-Pb system, the points for cogenetic minerals will lie in a straight line called the discordia. The upper point *A* where the concordia and discordia intersect is the true age of the mineral, while the lower point *D* of the intersection may be interpreted as the time of the metamorphism which led to the redistribution of the elements. This time does not always correspond to the actual geological age.

4.5.2 Lead Isotope Systematics

Variations in the isotopic compositions of lead can be used to study the petrogenesis of rocks, the genesis of sulphide ores, and the source and generation time of magma, etc. The common lead method is employed here, i.e. lead whose isotopic composition corresponded to a primaeval homogeneous Earth T years ago. The reservoir, in which U, Th, and Pb were uniformly distributed, formed soon after the Earth was formed. The differentiation of the mantle gradually led to the emergence of new reservoirs and nonuniform distribution of U/Pb and Th/Pb. As a result of the radioactive decay of U and Th the primaeval lead was in the course of time supplied with the radiogenic isotopes of ^{206}Pb, ^{207}Pb and ^{208}Pb. Moreover, the larger content of radioactive elements in the secondary reservoirs corresponded to greater supply of radiogenic lead. If at a moment t lead from this reservoir was extracted and redeposited in the form of lead minerals (for instance, galenite), then it would consist of two components: primary lead and radiogenic admixture, which accumulated within the time from T to t_1. If the other ore vein formed later at t_2, the common lead in this vein would consist of primary lead and radiogenic admixture whose content is higher than in the former case.

The isotopic composition of primary lead can be determined from meteoritic material, which is practically free of U and Th, by assuming that meteorites and the Solar System had a common origin. The primary composition is now taken to be that of the isotopic content of troilite from the meteorite in Diablo Canyon [4.13]:

$(^{206}Pb/^{204}Pb)_T = a_0 = 9.307$;
$(^{207}Pb/^{204}Pb)_T = b_0 = 10.294$;
$(^{208}Pb/^{204}Pb)_T = c_0 = 29.479$ the isotopic composition of lead at time T (primary lead), where T = 4.56 bln years ago

The evolution of isotopic composition of lead in time and is described by the integral equations

$$\left(^{206}\mathrm{Pb}/^{204}\mathrm{Pb}\right) = \left(^{206}\mathrm{Pb}/^{204}\mathrm{Pb}\right)_T + \lambda_{238}\int_T^t \mu(t)e^{\lambda_{238}t}\,dt \tag{4.18}$$

$$\left(^{207}\mathrm{Pb}/^{204}\mathrm{Pb}\right) = \left(^{207}\mathrm{Pb}/^{204}\mathrm{Pb}\right)_T + \frac{\lambda_{235}}{137.88}\int_T^t \mu(t)e^{\lambda_{235}t}\,dt \tag{4.19}$$

$$\left(^{208}\mathrm{Pb}/^{204}\mathrm{Pb}\right)_t = \left(^{208}\mathrm{Pb}/^{204}\mathrm{Pb}\right)_T + \lambda_{232}\int_T^t W(t)e^{\lambda_{232}t}\,dt \tag{4.20}$$

In order to use these equations it is necessary to know the forms of the functions $\mu(t)$ and $W(t)$, which are unknown. The simplest solution would be the assumption μ = const and W = const. It is called the single-stage model of lead evolution, the stage being the interval of time during which μ and W remain constant. The solution of these equations for this model is

$$\left(^{206}\mathrm{Pb}/^{204}\mathrm{Pb}\right)_t = a_0 + \mu\left(e^{\lambda_{238}T} - e^{\lambda_{238}t}\right) \tag{4.21}$$

$$\left({}^{207}\text{Pb}/{}^{204}\text{Pb}\right)_t = b_0 + \frac{\mu}{137.88}\left(e^{\lambda_{235}T} - e^{\lambda_{235}t}\right) \tag{4.22}$$

$$\left({}^{208}\text{Pb}/{}^{204}\text{Pb}\right)_t = c_0 + W\left(e^{\lambda_{232}T} - e^{\lambda_{232}t}\right) \tag{4.23}$$

By dividing the interval from T to t into n segments, we get an n-stage model. For instance, (4.22) for a two-stage model will change to

$$^{206}\text{Pb}/^{204}\text{Pb} = a_0 + \mu_1\left(e^{\lambda_{238}T} - e^{\lambda_{238}t_1}\right) + \mu_2\left(e^{\lambda_{238}t_1} - e^{\lambda_{238}t_2}\right) \tag{4.24}$$

In the equations (4.19)-(4.25), the following notation was used a_0 = $(^{206}\text{Pb}/^{204}\text{Pb})_T$; b_0 = $(^{207}\text{Pb}/^{204}\text{Pb})_T$; c_0 = $(^{208}\text{Pb}/^{204}\text{Pb})$; μ = $^{238}\text{U}/^{206}\text{Pb}$; W = $^{232}\text{Th}/^{208}\text{Pb}$; $W = \mu K$; where K is $^{232}\text{Th}/^{238}\text{U}$, t is the time at the moment of measurement; T is the age of the Earth, 137.88 = $^{238}\text{U}/^{235}\text{U}$, and λ_{238}, λ_{235} and λ_{232} are the decay constants of ^{238}U, ^{235}U and ^{232}Th, respectively.

It is impossible to use (4.22), (4.23), and (4.24) to estimate the age of individual lead ore deposits since each contains two unknowns, viz. t and μ (or W). By solving (4.22) and (4.21) jointly we can exclude U and get the equation for a single-stage isochrone.

Figure 4.10 contains the evolution curves for the isotopic ratios of Pb with given μ and W. The intersection of the isochrones and growth curves for different μ yields the growth. This is the one-stage model of Holmes-Houtermanns (Fig. 4.11). All the isochrones originate at the same point, corresponding to a_0, b_0, c_0. The isochrone corresponding to zero-age is called the Geochrone. It serves as the location of contemporary lead, for instance, extracted from oceanic basalts. The isotopic composition of lead will yield the age t (the evolution time of lead), μ ($^{238}\text{U}/^{206}\text{Pb}$) and K ($^{232}\text{Th}/^{238}\text{U}$) for the regional source of lead. The following data was given by Aswathanarayana [4.11] from the galenites of Mamandar in India for $(^{206}\text{Pb}/^{204}\text{Pb})_t$ = 14.40; $(^{207}\text{Pb}/^{204}\text{Pb})$ =15.12; $(^{208}\text{Pb}/^{204}\text{Pb})$ = 34.31.

Using these values of a_0 and b_0, the slope of the isochrone (m) can be calculated:

$$m = \frac{\left(^{207}\text{Pb}/^{204}\text{Pb}\right) - b_0}{\left(^{206}\text{Pb}/^{204}\text{Pb}\right) - a_0} = \frac{15.12 - 10.294}{14.40 - 9.307} = 0.9478$$

From tables (Appendix 3), the age of the isochrone corresponds to the interval of 2.2-2.4 bln years. Linear interpolation gives us 2260 mln years.

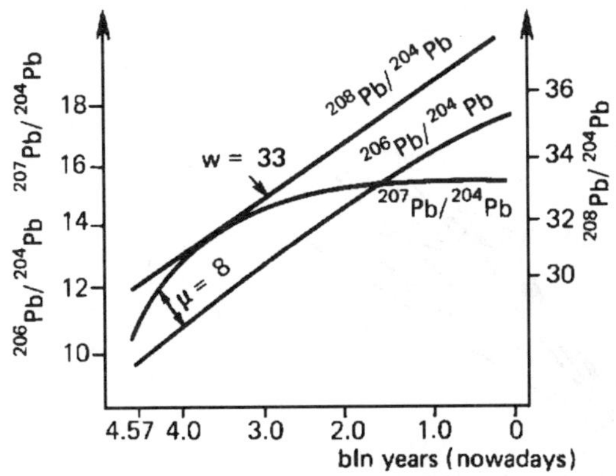

Fig. 4.10. Evolution of isotopic composition of lead (according to Delaoi, 1980)

From this age, we can calculate μ and *K*:

$$\mu = \frac{\left(^{206}\mathrm{Pb}/^{204}\mathrm{Pb}\right) - a_0}{e^{\lambda_{238}} - e^{\lambda_{238}t}} = \frac{14.40 - 9.307}{2.0255 - 1.4202} = 8.401$$

$$K = \frac{^{232}\mathrm{Th}/^{204}\mathrm{Pb}}{^{238}\mathrm{U}/^{204}\mathrm{Pb}} = \frac{34.31}{8.41} = 4.08 \qquad (4.25)$$

The systematics of lead isotopic composition is used to study the evolution and history of the geospheres, the genesis of lead ore deposits, etc. Various models are used that can be divided into two groups, viz. the evolution and mixture models.

In research on the evolution of the isotopic composition of lead in crustal and mantle matter, the model objects used to construct lead-isotopic systematics are usually tholeiitic basalts of mid-oceanic ridges, to provide information on the depleted mantle (MORB), oceanic islands (the mantle under hot beds, OIB), and lead ores with diverse geneses.

An example is the popular model of B. Doe and R. Zartman [4.14], called plumbotectonics. This model is based on several thousand isotopic analyses of lead and indicates that we should not expect any overlap of fields for the basalts of mid-oceanic ridges, the interplate hot beds and pelagic sediments, which shows the absence of a genetic bonds between them in the diagrams of ^{206}Pb/^{204}Pb-^{207}Pb/^{204}Pb (α-β) and ^{206}Pb/^{204}Pb-^{208}Pb/^{204}Pb (α-γ). Oceanic basalts have a ^{207}Pb deficit, which proves the antiquity of the processes which cause

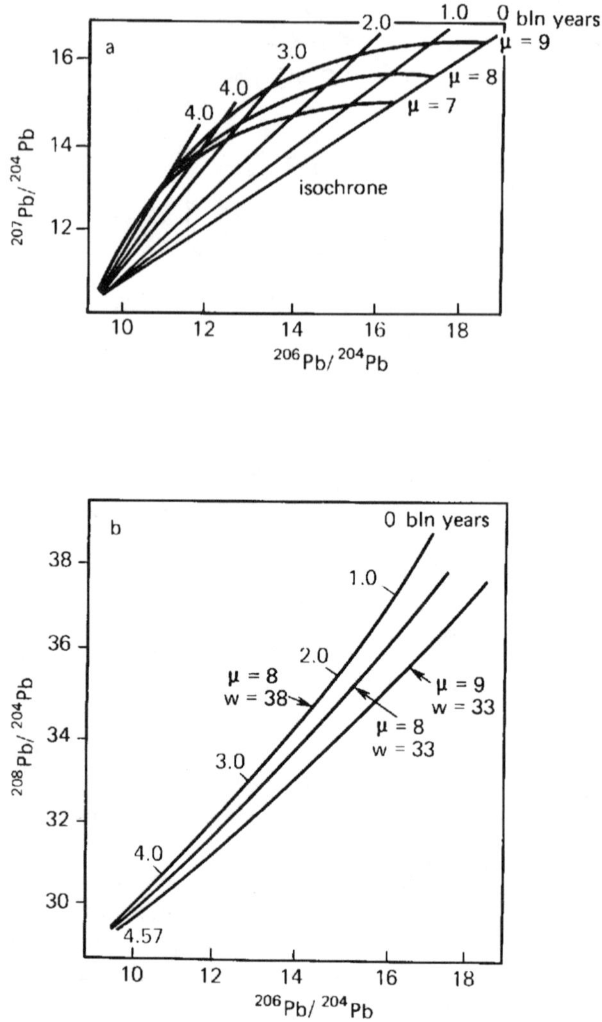

Fig. 4.11. Evolution of lines of lead (according to M. Delaoi, 1980) in coordinates:
(a) $^{207}Pb/^{204}Pb$–$^{206}Pb/^{204}Pb$
(b) $^{208}Pb/^{204}Pb$–$^{206}Pb/^{204}Pb$

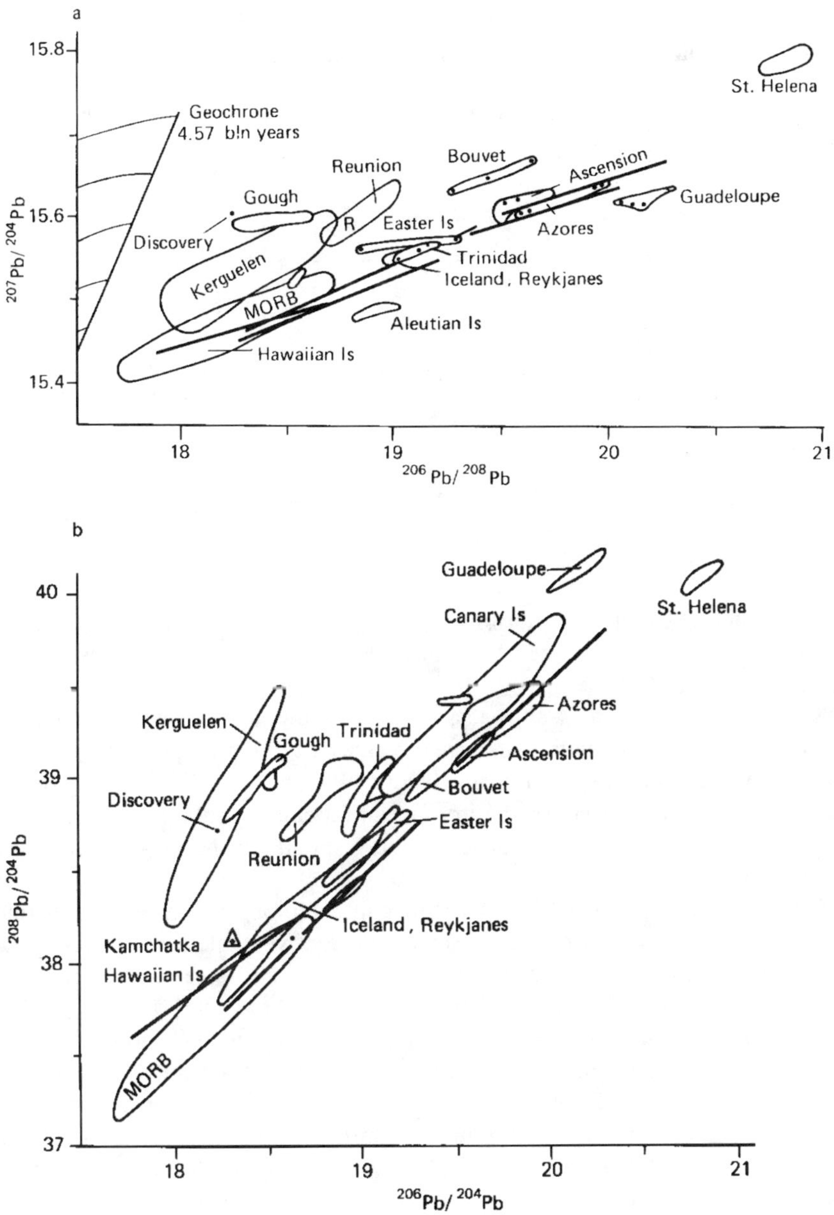

Fig. 4.12. Diagram of isotopic relations of lead for the volcanic rocks of oceanic islands, in coordinates:
(a) $^{207}Pb/^{204}Pb$–$^{206}Pb/^{204}Pb$; (b) $^{208}Pb/^{204}Pb$–$^{206}Pb/^{204}Pb$ (according to S.-S. Sun, 1980)

the isotopic composition of lead in the mantle source of these basalts. The lead in primitive island arcs is close to MORB lead, while the lead composition in mature islands has much in common with pelagic sediments, i.e. with continental lead (Fig. 4.12) [4.15].

Jointly solving Eqs. (4.22) and (4.23) yields the original value of K (Th/U at T) for a single-stage model:

$$K = \frac{\left({}^{208}\mathrm{Pb}/{}^{204}\mathrm{Pb}\right)_t - c_0}{\left({}^{206}\mathrm{Pb}/{}^{204}\mathrm{Pb}\right)_t - a_0} \cdot \frac{\left(e^{\lambda_{232}T} - e^{\lambda_{232}t}\right)}{\left(e^{\lambda_{238}T} - e^{\lambda_{238}t}\right)} \tag{4.26}$$

The model proves fairly tenable for contemporary oceanic basalts. For $t = 0$, (4.27) is simplified to

$$K = \frac{\left({}^{208}\mathrm{Pb}/{}^{204}\mathrm{Pb}\right)_t - c_0}{\left({}^{206}\mathrm{Pb}/{}^{204}\mathrm{Pb}\right)_t - a_0} \cdot \frac{\left(e^{\lambda_{232}T} - 1\right)}{\left(e^{\lambda_{238}T} - 1\right)} \tag{4.27}$$

The systematics of the thorium-uranium ratio 'K' in primary source of volcanic rocks is characterized by a distinct distribution pattern. Two factors must be taken into account.

(1) The value K from (4.28) or (4.27) is an integral value which takes into consideration only the initial and final values of the isotopic ratios.

(2) The variations in the isotopic ^{208}Pb/^{204}Pb and ^{206}Pb/^{204}Pb ratios during the time of existence of contemporary Ocean (about 160 mln years) are within the limits of analytical error, that is, K_{Pb} corresponds to the pre-oceanic phase of the source development.

The value of K_{Pb} = (Th/U)$_{Pb}$ [4.16] in contemporary basalts has a distinct distribution (Fig. 4.13). Most samples from current volcanic rocks of the ocean floor, oceanic islands, primitive island arcs or primitive series in developed island arcs are characterized by a constant value of $K_{Pb} = 3.8 \pm 0.2$, independent of their petrochemical composition. The lowest K_{Pb} is only found in grey gneisses from the ancient cores of the continents: 3.56 (Amitsok, Greenland, 3.7 bln years) and 3.72 (Wyoming, USA, 2.6 bln years), which correspond to the primitive, weakly differentiated mantle. This means that the above-mentioned oceanic rocks are the mantle derivatives. In addition, the process of depletion associated with the melting of continental crust did not influence the Th/U ratio. Values of K_{Pb} higher than the primary ones may be due to contamination by sialitic material in the sediments of sea water (in hydration) during subduction and recycling. The isotopic composition of volcanic rocks

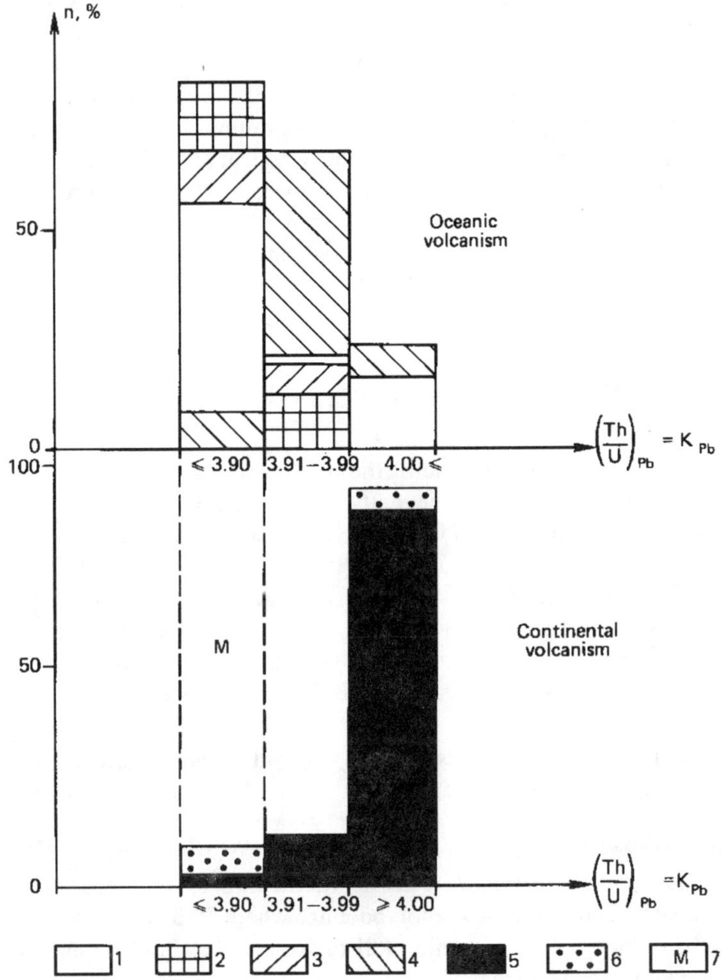

Fig. 4.13. Histogram of K_{Pb} distribution in the volcanic rocks of the world: *1*–MORB; *2*–island arcs; *3*–Iceland; *4*–oceanic islands; *5*–continental reefs; *6*–oceanic sediments; *7*–mantle zone; n–frequency distribution

of continents is heterogeneous and usually exceeds 4. Some K_{Pb} lower than oceanic values are observed in granites and are presumably due to the remelting of the primaeval non-differentiated material. The ocean today contains areas, where the K_{Pb} of volcanic rocks is very near to the continental values (≥4.0) and the oceanic sediments (4.15 on the average). Such a region lies in the large block in the South Atlantic Ocean between 34° and 50° South and contains part of the mid-atlantic ridge, the islands of Tristan-da Cunha, Gough, and the underwater mountain Discovery. Analogous values close to continental ones are shared by the islands near the continents (the Canary Islands, Guadeloupe), as well as a considerable part of the Indian Ocean. The proximity of K_{Pb}, the isotopic compositions of Sr, Th, and noncoherent elements in these zones that are inconsistent with continental values indicates they are traces of ancient continents.

4.6 The Rhenium-Osmium Method

This method makes use of the beta decay of ^{187}Re, the main isotope of natural rhenium into ^{187}Os. The half-life of ^{187}Re = 4.3 × 10^{10} years. The isotopic composition of natural rhenium and osmium is [4.4]:

^{185}Re = 37.07%	^{184}Os = 0.018%	^{188}Os = 13.3%
^{187}Re = 62.93%	^{186}Os = 1.59%	^{189}Os = 16.1%
	^{187}Os = 1.64%	^{190}Os = 26.4%
		^{192}Os = 41.0%

According to Goldsmith's classification, Re and Os are siderophile elements, whence their distribution and application. Re is a Group VIIA element and a chemical analogue of Mo ($R_{Re^{4+}} = 0.72$ Å, $R_{Mo^{4+}} = 0.74$ Å). The average content of Re in the earth's crust is 7 × 10^{-8}% (according to A. Beus, 1975). The principal mineral containing it is molybdenite, where ReS_2 is isostructural with MoS_2. The similarity between ionic radii of Re and Ta (0.69 Å) and Nb (0.68 Å) may explain the presence of rhenium in the minerals of these elements.

Osmium belongs to the platinum group (VIIIA). Its average content in the earth's crust is about 10^{-7}%. Osmium is practically only found in ultrabasic rocks, where it associates with iridium. The content of Re in iron meteorites is of the order of 10^{-6}%, while that of Os is 10^{-5}%.

The method is mainly employed to date iron meteorites and molybdenum deposits [4.1, 4.18]. Age is estimated by the method of isochrones using the equation

$$\frac{^{187}\mathrm{Os}}{^{186}\mathrm{Os}} = \left(\frac{^{187}\mathrm{Os}}{^{186}\mathrm{Os}}\right)_0 + \frac{^{187}\mathrm{Re}}{^{186}\mathrm{Os}}\left(e^{\lambda_{187}t} - 1\right) \tag{4.28}$$

The primary ratio $^{187}Os/^{186}Os$ = 0.831 was established in [4.1].

The method is not widely applied because of the low concentrations of Re and Os and the long half-life of ^{187}Re and is basically used in cosmic research.

4.7 The Lutecium-Hafnium Method

The method is based on the beta decay of ^{176}Lu with a half-life of 3.5 × 10^{10} years to yield ^{176}Hf. The method is poorly developed because of difficulties in establishing precise isotopic composition of hafnium. Lutecium belongs to the group of heavy rare-earth elements and has the smallest ionic radius among the trivalent lanthanides (0.85 Å).

It is concentrated in heavy lanthanide minerals, such as xenotime and fergusonite. Presumably Lu^{3+} is capable of heterovalent isomorphism with Ca^{2+} (0.99 Å) in mafic rock-forming minerals. Lutecium is not widespread with lanthanides. Its content in basalts is on average 0.46 ppm and 0.45 ppm in granites. The average content in the continental lithosphere is 0.8 ppm according to A. Beus (1975).

Hafnium is Group IV element in the Periodic Table and is a chemical analogue of zirconium ($R_{\mathrm{Hf}^{4+}} = 0.78$ Å, $R_{\mathrm{Zr}^{4+}} = 0.74$ Å). Its average content in the continental lithosphere is 2.8 ppm (A. Beus, 1975), which is 50 times less than the content of zirconium.

In this respect hafnium is a dispersed element, which isomorphously replaces zirconium in zirconium minerals. It is principally concentrated in zircons. Zircons are unsuitable for geochronological analysis, since they contain large quantities of common hafnium. Apatites, garnets, monazites, etc. are usually used for analysis.

The crystallization age of minerals is established by the isochrone equation:

$$\frac{^{176}\mathrm{Hf}}{^{177}\mathrm{Hf}} = \left(\frac{^{176}\mathrm{Hf}}{^{177}\mathrm{Hf}}\right)_0 + \frac{^{176}\mathrm{Lu}}{^{177}\mathrm{Hf}}\left(e^{\lambda_{176}t} - 1\right) \tag{4.29}$$

At present there are only preliminary data on hafnium isotopic systematics for research into global reservoirs, which share features typical of strontium and neodymium systematics [4.1].

REFERENCES

4.1. Faure, G., 1986. Principles of Isotope Geology, Second Edition. John Wiley & Sons, Inc. New York.

4.2. Dalmeyer, R.D., Lamper, M.A., 1969. Potassium-argon Dating, W.H. Freeman and Co. San-Francisco.

4.3. Lectures in Isotopic Geology, 1979. (Ed. by E. Jaeger and J.C. Hunziker). Springer Verlag, Berlin-Heidelberg, New York.

4.4. Hamilton, E.I., 1965. Applied Geochronology, Acad. Press, London, New York.

4.5. Tolstikhin, I.N., 1960. Isotopic Geochemistry of Helium, Argon and Rare Gases, Nauka, Leningrad.

4.6. The Evolution of the Crust-Mantle, 1986. (Ed. by Yu.A. Shukolyukov). Nauka, Moscow.

4.7. Balashov, Yu.A., 1985. Isotopic and Geochemical Evolution of Mantle and the Earth Crust, Nauka, Moscow.

4.8. De Paolo, D.J. Wasserburg, G.J., 1977. The Sources of Island Arcs as Indicated by Nd and Sr Isotopic Studies.—Geophys. Res. Let., **4**, No 10, 465-468.

4.9. White, W.M., Hofmann, A.W., 1982. Sr and Nd Isotopic Geochemistry of Oceanic Basalts and Mantle Evolution.—Nature, 296, 821-825.

4.10. Awwiller, D.N., Mock, L.E., 1991. Diagenetic Modification of Sm-Nd Model Ages in Tertiary Sandstones and Shales Texas Gulf Coast.—Geology, **19**, No 4.

4.11. Aswathanarayana, V., 1985. Principles of Nuclear Geology. Oxonian Press Pvt. Ltd, New Delhi.

4.12. Tugarinov, A.I., 1973. General Geochemistry. Atomizdat, Moscow.

4.13. Tatsumoto, M., Knight, R.J., Allégre, C.J., 1973. Time Difference in the Formation of Meteorites as Determined from the Ratio of Lead 207 to Lead 206. *Science*, **180**, 1279-1283.

4.14. Doe, B.R., Zartman, R.E., 1979. Plumbotectonics-1, the Phanerozoic.—Geochemistry of Hydrothermal Ore Deposits. John Wiley & Sons, Inc. New York.

4.15. Sun S.-S., 1980. Lead Isotopic Study of Young Volcanic Rock from Mid-Ocean Ridges, Ocean Islands and Island Arcs.—Phil. Tran. Roy. Soc., London, **A297**, No 1430, 409-445.

4.16. Titayeva, N.A., 1990. Geochemical Tracer Role of Isotopic Values of Thorium-Uranium Ratio in Volcanic Rocks. USSR—Acad. Sc. Reports., **312**, No 6, 1454-1457.

4.17. Walker, R.J., Morgan, J.W., Naldertt, A.J., Li, C. and Fasset, J.D., 1991. Re–Os Isotope Systematics of Ni–Cu Sulfide Ores, Sudbury Igneous Complex, Ontario: Evidence for Major Crustal Component.—Earth and Planet. Sci. Lett., **105**, No 4, 416-429.

5 Radionuclides in Natural Decay Series. Disequilibrium Isotopic Method

5.1 The Decay Series

As distinct from other long-lived natural radionuclides, ^{238}U, ^{235}U, and ^{232}Th do not immediately decay into stable products, but form numerous relatively short-lived intermediates, thus yielding decay series or radioactive families (App. 1). The final stable product of all three series are lead isotopes, which were considered in the preceding chapter. The decay series were discovered by the Curies, who found that the radioactivity of uranium ores (containing all the decay products in addition to uranium) is several times higher that of the pure salts of uranium (comprising only the isotopes of uranium). In the decay products, new chemical elements, viz. radium, polonium, radon, actinium, and protactinium were discovered. Since at that time the phenomenon of isotopy was unknown, the newly discovered radioactive substances were given new names, and considered as independent chemical elements. It was only after Soddy introduced the concept of isotopy that it became apparent that the radioactive decay products were isotopes of known elements. For this reason, each term in the series has two indices: the old one showing the link with the decay series, and the common nuclide symbols. For instance, the beginning of the ^{238}U decay series is written as follows

$$^{238}_{92}U(UI) \xrightarrow{\alpha} {}^{234}_{90}Th(UX_1) \xrightarrow{\beta} {}^{234}_{91}Pa(UX_2) \xrightarrow{\beta}$$
$$^{234}_{92}U(UII) \xrightarrow{\alpha} {}^{230}_{90}Th(Io) \xrightarrow{\alpha} {}^{226}_{88}Ra \xrightarrow{\alpha} {}^{222}_{86}Rn$$

Radioactive decay is an irreversible phenomenon.Therefore from the beginning of the series to its end the ordinal numbers of elements usually decrease as do the mass numbers when there is alpha decay. (In beta decay the atomic number is increased by unit while the mass number remains the same.) Since the emission of an alpha particle reduces the mass number by 4, there are, in principle, four types of series, where the mass numbers are $4n$, $4n+1$, $4n+2$, $4n+3$ (n being an arbitrary nonzero number). If we consider the mass numbers of natural radioactive series, we can see that they follow this pattern. For instance, the ^{232}Th series corresponds to the $4n$ type (232 = 4 x 58); the ^{238}U series corresponds to ($4n$ + 2) type (238 = 4 x 59 + 2); and the ^{235}U series corresponds to ($4n$ + 3) type (235 = 4 x 58 + 3). The ($4n$ + 1) series has not been found in nature, although members of this series have been synthesized by artificial nuclear reactions. This is the 'neptunium' series called after its most long-lived member $^{237}_{93}$Np, which has a half-life of 2.25×10^6 years. The final stable product of the series is $^{205}_{81}$Tl. It is thought that the natural element $^{209}_{83}$Bi ($T_{1/2} = 2.7 \times 10^{17}$ years) is the only survived member of the series. The rest members decayed during the early history of the Earth. The discovery of new artificial radionuclides makes it possible to extend decay series thus arriving at its beginning. An ancestor of ^{238}U is ^{258}Md, for ^{232}Th it is ^{256}Es, for ^{235}U, ^{247}Cm, for ^{237}Np, ^{241}Pu and so on. Moreover, secondary branches of the series have been found. The principal and secondary branches have different parents but coincide beginning with a certain common member. All radioactive isotopes are distributed among the four decay series.

In each natural series there is a transformation sequence, i.e. one alpha decay is followed by two beta decays, or vice versa. Alpha decay decreases the charge of the nucleus by two units, while two beta decays increase it by two units. The result is a new isotope of the original element, whose mass number four units less. An example is the ^{234}U isotope from ^{238}U and ^{230}Th from ^{234}Th in the ^{238}U series partially given above.

Natural series have much in common. They all end in stable isotopes of lead. In the middle of each series we find an isotope of the only radioactive gas radon. Radon divides the series into two parts. The first part contains the most long-lived radionuclides, isotopes of U, Th, Pa, Ac, Ra, and Fr. The rest parts of all the three series are similar even by their configurations, viz. the short-lived isotopes of Pb, Bi, Po, Tl, and At. Many of them can disintegrate in two ways (α and β), giving rise to 'bifurcation', which then converges to the same radionuclide.

The radionuclides in the end part of these series have the highest energies of radiation decay. The maximum γ-radiation energy in the ^{238}U series is radiated by ^{214}Bi (1.76 MeV), while in the ^{232}Th series, it is ^{212}Tl (2.62 MeV). The latter has the highest γ-radiation energy among all natural radionuclides. Similarly, ^{212}Po has the maximum alpha radiation energy (10.5 MeV) among all natural radionuclides.

The neptunium series does not resemble the other three series. We only deal with natural decay series in this chapter.

5.2 The Geochemistry of Radioactive Elements in Decay Series

5.2.1 Radium

Four isotopes exist in nature, viz. ^{224}Ra ($T_{1/2}$ = 3.6 days), ^{226}Ra ($T_{1/2}$ = 1602 years), ^{228}Ra ($T_{1/2}$ = 5.9 years); and ^{223}Ra ($T_{1/2}$ = 11.2 days). Radium is the heaviest element in group IIA elements. It is a chemical analogue of Ba, viz., R_i (Ba^{2+}) = 1.38 Å; R_i (Ra^{2+}) = 1.44 Å. Most of the salts of radium are isostructural with barium salts. Ra is more basic than Ba. Radium salts are much less soluble than the corresponding Ba salts. As is the case with all alkaline-earth elements, Ra has one oxidation number +2 and does not tend to form complexes. The radium salts $RaCl_2$, $RaBr_2$, RaI_2, $Ra(NO_3)_2$, and RaS are soluble in water, while $RaSO_4$, RaC_2O_4, $RaCO_3$, $Ra_3(PO_4)_2$, $RaCrO_4$, $Ra(IO_3)_2$, and RaF_2 are practically insoluble in water. Radium sulphate, carbonate and chromate are used to separate radium together with Ba as a carrier from solutions.

The Behaviour of Radium in Nature

The Geochemistry of radium is affected by (1) its chemistry, the most important properties are its chemical similarity to Ba, the high solubility of the chlorides, poor solubility of the sulphate and carbonate, its ionic form in solution; and (2) the relatively short half-life of radium isotopes, which means that rapidly processes commensurable with their half-lives, they will be considered as isotopes of independent chemical element Ra. In slower processes radium atoms, no longer maintained by parent radionuclides, decay. A radioactive equilibrium is established in a decay series and the presence of a radium isotope will only depend on the presence of a parent isotope of uranium or ^{232}Th. Therefore, the main independently existing isotope of radium is the most long-lived ^{226}Ra (^{238}U series). Unsupported atoms of ^{226}Ra will completely decay in over 10 thousand years. The behaviour of ^{228}Ra (^{232}Th series) can also be readily analysed since it exists for about 40 years. Other isotopes of radium lack geochemical significance due to their short half-lives. The independent behaviour of the radium isotopes can be used in dynamic processes which take place at present in natural waters and biological processes.

3. Radium isotopes occur at very low concentrations in nature. The average content of the most widespread isotope ^{226}Ra in the earth's crust is about 10^{-10}%, and in natural waters between $n \cdot 10^{-14}$ g/l and $n \cdot 10^{-12}$ g/l (with the exception of Cl—Na—Ca brines in the oil or gas fields and water in contact with uranium ores). Its low concentration in natural water determines the

behaviour of radium atoms characteristic of radionuclides in dilute solutions (Ch. 2). In the presence of a carrier (Ba) radium behaves in compliance with its geochemical properties. If the concentrations of Ba in the water is infinitesimal, the radium atoms behave as typical radiocolloids. They can either absorb on any negatively charged solid or coprecipitate with metals hydroxides.

The behaviour of radium in magmatic process has not received much attention. It is assumed that at every stages it follows barium and accumulates in the residual melt. Uranium and thorium are lithophilic, and hence the residual melt is rich in uranium and other relatively long-lived decay products. The radium distribution coefficients between rock-forming minerals and the coexisting melt are not yet known, and so we have no knowledge concerning the relationship between uranium and radium during magmatic melt evolution. In magmatic rocks older than 10 thousand years, ^{238}U and ^{226}Ra will be in radioactive equilibrium unless exogenous processes change their ratio. During volcanic eruptions radium can be released as a chloride in the gas phase. Contemporary volcanic rocks are often richer in ^{226}Ra than ^{238}U, which under these conditions is less volatile, while in the volcanic minerals there are barium chlorides.

The behaviour of radium in hydrothermal process can be traced using current hydrotherms. The behaviour of radium is significantly influenced by the oxidation number of sulphur. Radium enters hydrothermal solutions in zones of active volcanism together with the volcanic gases in the form $Ra(Ba)Cl_2$. At the oxidizing barrier within the zone where sulphates appear, $Ra(Ba)SO_4$ precipitates in pores and cracks of volcanic rocks. This means that water coming to the surface contain low concentrations of radium, whereas in the zone where barium sulphate precipitates, the deposit is rich with radium. The existence of deposits of this kind at relatively small depths can be seen from the high concentrations of radon at the exit of contemporary hydrotherms, the concentrations being sometimes hundreds and even thousands times greater than the background, in spite of the low concentrations of uranium in the rocks [5.1].

The exogenic geochemistry of radium has been studied in much greater detail. Radium, as a recoil atom, can easily be leached from the rocks by natural water. In sulphateless fissure waters of granitic massifs, which do not contain colloids, the concentration of radium exceeds its concentration in surface waters by an order or more [5.2]. Zones where fissure and ground waters are in contact, become enriched with colloidal particles and there is an increase in the pH and the sorption of Ra on the walls of water-conducting fractures. Hence, fissure rocks near the surface are always richer in radium over equilibrium value with parent uranium that forms more stable compounds in aqueous solutions. Similarly, bottom silty deposits are richer in radium where fissure water discharges.

Cl–Na–Ca brines in contact with hydrocarbon deposits have a special place in the geochemistry of radium. The medium promotes the reduction of sulphates. A large number of metals, including Ba and Ra are leached out by highly mineralized chloride brines from water-enclosing rocks. The content of Ra in such waters is 100-1000 times higher than in sulphate waters beyond the deposit [5.3]. The greater radioactivity of water where there are water-oil or water-gas interfaces is an indicator of oil or gas fields. The biochemistry of radium is interesting. In acid soils typical of humid zones, the ^{226}Ra concentration usually exceeds the equilibrium concentration with ^{238}U. The radium is correlated with clay minerals, viz., Fe, Mg and Al hydroxides and organic matter, which indicates its sorption and ion-exchange origin. In black soils and soils in arid climates radium accumulates on carbonate and sulphate evaporating barriers, by coprecipitating with the carbonates and sulphates of other alkaline-earth metals.

The coefficient of biological absorption (CBA) of radium is greater than unit. This makes it a non-barrier element. The highest coefficients are found in the plants of humid climates. The CBAs in arid climates are lower because the root extractions of plants easily recover either the exchange forms of radium or forms sorbed by solid surfaces. In arid climates radium is bound in poorly soluble barium sulphate or calcium humates [5.4].

5.2.2 Radon

Natural radon has three isotopes: ^{222}Rn (radon), ^{220}Rn, and ^{219}Rn, which are respectively in the ^{238}U, ^{232}Th, and ^{235}U decay series. The isotopes of radon have rather short half-lives, viz., 3.8 days for ^{222}Rn, 54.5 s for ^{220}Rn, and 3.9 s for ^{219}Rn, which determines their geochemistry.

The chemistry of radon is typical of a noble gas. It does not react with oxygen even in a spark discharge or in the presence of catalysts. It is chemically inert and its valency is zero. However, Rn can be found in clathrate compounds with water, phenol, and toluene. In clathrate compounds the bond is due to van der Waals forces. Rn combines with gaseous fluorine to yield RnF_4 and co-crystallizes with xenon. Like krypton and xenon, Rn forms hexahydrates. $Rn \cdot 6H_2O$ is isomorphous with $H_2S \cdot 6H_2O$ and $SO_2 \cdot 6H_2O$. In normal conditions radon is found to be in a monatomic form.

Radon isotopes are soluble in water and other liquids. The solubility coefficient in H_2O at 15 °C varies from 0.25-0.30. The solubility of radon drops with increasing temperature while boiling removes it completely from water.

The solubility of radon in organic liquids is considerably higher. Its solubility coefficients at 18 °C are [5.5]: 6.17 for ethyl alcohol, 12.82 for benzene; 15.08 for ether; 29.0 for olive oil; 10.00 for vaseline and kerosene. Its solubility in fats means that it can concentrate in human fat tissue, which must be taken into consideration when assessing radiation hazards.

Rn is sorbed by solids. Various inorganic gels and organic colloids firmly retain adsorbed radon. The best sorbent is activated carbon. Adsorbed radon is mobile and readily redistributes in solid from surface molecular layers into deeper layer. Radon is desorbed upon heating. It is completely desorbed from activated carbon at 350-400 °C. The evolve of radon from one phase to another is called emanation (a term is derived from the old name of radon as the emanation of radium).

The emanation coefficient k_{Rn} is defined as the ratio of the amount of radon liberated from a solid or liquid body to the quantity formed in the body during the same interval of time. The value of k_{Rn} depends on the crystalline structure of the mineral, porosity, and fracture of the rocks. It fluctuates from fractions of a percent (e.g. monazite) to tens per cent (e.g. soil, coal, uranium ore, minerals of weathered crusts). For plants k_{Rn} reaches 100% [5.5].

The short-lived isotopes of radon can exist for relatively short periods of time ($^{222}Rn \approx 27$ days, $^{220}Rn \approx 5$ min, $^{219}Rn \approx 0.5$ min). They are constantly accumulated in objects containing the parent isotopes of radium, or uranium and thorium minerals. As a result of emanation radon isotopes fill pores, capillaries and fissures of rocks. The concentration of radon C_{Rn} in a subterranean atmosphere can be estimated as follows [5.3]

$$C_{Rn} = ak_{Rn}Q_{Ra} \cdot \rho \frac{1000}{\eta} \text{ Ci} / \text{l} \quad (5.1)$$

where k is the emanation coefficient, %, ρ is the density of rock, g/cm^3, η is the porosity of rock, %; a is a proportionality factor (for ^{222}Rn $a = 1$, for ^{220}Rn $a = 1.6 \cdot 10^5$, and for ^{219}Rn $a = 5.1 \cdot 10^4$), Q_{Ra} is the content of the parent radium isotope, C/g of rock. The value of a is the amount of Rn isotope formed per cm^3 of rock per s. The value of a can be calculated from the proportion

$$\left(\frac{a}{\lambda}\right)_{222} \div \left(\frac{a}{\lambda}\right)_{220} \div \left(\frac{a}{\lambda}\right)_{219}$$

Radon migrates in the earth's crust as a gas or in a dissolved state. The content of radon in a subterranean atmosphere and the hydrosphere depends on the concentration of radium in the enclosing rocks and its emanation coefficient. Friable or fissured rocks, which have the large internal surfaces, have large emanation coefficient (tectonic dislocations, weathered crust, etc.). Above uranium or thorium deposits we find gas halo dispersion. The extension

Table 5.1 The Content of ^{226}Ra and ^{222}Rn in Natural Waters [5.2]

Type of water	Natural situation	^{226}Ra, g/l	^{222}Rn, Em*
Surface waters	Oceans	8×10^{-14}-4.5×10^{-11}	0
	Seas	$(0.1\text{-}8)\times10^{-12}$	0
	Rivers	$(0.25\text{-}4.2)\times10^{-12}$	0
Waters of sedimentary rocks	Zone of intense water exchange	$(1\text{-}6)\times10^{-12}$	1.0-50.0
	Zone of hampered water exchange	(1×10^{-11})-1×10^{-8}	1.0-20.0
Waters of acid magmatic rocks	Waters of weathering crusts	$(1\text{-}7)\times10^{-12}$	10.0-400.0
	Waters of deep-seated tectonic fissures	$(2\text{-}9)\times10^{-12}$	8.0-400.0
Waters of uranium deposits	Waters of oxidation	8×10^{-12}-2×10^{-9}	50.0-5.10^{4}
	Waters of reduction zone	$(0.1\text{-}8.0)\times10^{-10}$	50.0-3×10^{3}

* 1 Em = $1\cdot10^{-10}$ Ci/l

of the halo dispersion depends on the permeability of rocks and the half-life of the radon isotope. The largest gas halo dispersion (tens of metres) above uranium ores contains the relatively long-lived ^{222}Rn. Halo dispersion of ^{220}Rn and ^{219}Rn is small and ^{219}Rn halo dispersion always accompanies ^{222}Rn halo dispersion. The solubility of radon in water ensures its migration with underground water, which increases the extent of radon halo dispersion. Water under pressure in fractures draining from fissured rocks invariably contain high concentrations of radon, sometimes as much as a hundred pC/l. The distribution of ^{222}Rn and ^{226}Ra in natural water is given in Table 5.1.

The highest concentrations of ^{222}Rn occur at the exits of underground water; they are substantially lower in sea water (from 0.2 to 9 pC/l). Waters from deep tectonic fissures in the Alpine orogenesis often have high concentrations of radon and have some balneological interest (radon nitrogen-containing thermal waters, radon carbonate waters). In a number of cases spring water is rich in radon due to near-surface accumulation of radium at sulphate or carbonaceous barriers.

Table 5.2 The Change in the Content of the Isotopes of Radon in the Atmosphere [5.2]

Height above the Earth's surface, m	0.01	1	10	100	1 000	7 000
Relative content of ^{222}Rn, %	100	95	87	69	38	7
Height above the Earth's surface, m	0	5	10	25	50	100
Relative content of ^{220}Rn, %	100	70	50	29	5	0.5

The difference between the concentrations of radon in the soil air and the atmosphere causes a continuous convection of radon into the atmosphere. The concentration of radon isotopes decreases away from the surface of the Earth (Table 5.2) with the ^{220}Rn concentration decreasing more rapidly than that of ^{222}Rn.

The short-lived decay products of radon isotopes are formed in the atmosphere in the form of free atoms and are rapidly transformed into positive ions as a result of the ionization by alpha particles. These ions stick to aerosol particles or water droplets and fall with them to the Earth's surface.

5.2.3 Polonium

The natural isotopes of polonium are situated at the end of the natural radioactive series and have short half-lives:

^{210}Po (138 days), ^{214}Po ($1.6 \cdot 10^{-4}$ s), ^{218}Po (3.05 min) — the series ^{238}U

^{211}Po (0.52 s) and ^{215}Po ($1.8 \cdot 10^{-3}$ s) — the series ^{235}U

^{212}Po ($2.9 \cdot 10^{-7}$ s) and ^{216}Po (0.15 s) — the series ^{232}Th

The most long-lived isotope of polonium is ^{210}Po and is used in geochemistry (biogeochemistry).

The chemical properties of polonium are those of Group VIA. It has the same electronic structure as selenium and tellurium, and approaches the latter in terms of its chemistry. As a group VIA element it has several oxidation members: -2, +2, +3, +4, and +6. Po(III) has much in common with Bi. Po(IV) is the most stable in solutions. Polonium is present in natural solutions in ultra-microconcentrations. It forms pseudocolloids, which emerge as a result of the adsorption of the ionic forms of Po on the colloidal particles of silicon dioxide, organic colloids, etc. The adsorption has an ion-exchange character. At pH of 1-4 Po exists in nonhydrolyzed ionic forms; at pH of 6-7 hydrolysis occurs and positively charged colloidal particles form; in the case of pH 8-9 it exists as negatively charged colloidal hydroxide particles. In a strongly alkaline medium the anion PoO_3^{2-} is formed. Pseudocolloids form in solutions where solubility product cannot be achieved.

Po forms complexes with diverse anions. Po is very toxic. Only the most long-lived isotope, ^{210}Po, has geochemical significance. It is formed by the decay of ^{222}Rn. Its usual associate is ^{210}Pb (22 years) from which ^{210}Po derives in the chain:

$$^{210}\text{Pb} \underset{(22\ \text{years})}{\overset{\beta}{\rightarrow}} {}^{210}\text{Bi} \underset{(5\ \text{days})}{\overset{\beta}{\rightarrow}} {}^{210}\text{Po} \underset{(138\ \text{days})}{\overset{\alpha}{\rightarrow}} {}^{206}\text{Pb}$$

In the atmosphere ^{210}Po and ^{210}Pb, which form from ^{222}Rn, are rapidly sorbed by aerosol particles. The radionuclides enter the soils from the atmosphere as a result of 'dry' precipitation and with atmospheric precipitates. The ^{210}Pb concentration in rain and snow is between 0.1 and 16.2 pC/l, while that of ^{210}Po is from 0.2 to 1.36 pCi/l [5.3]. According to V.D. Vilenskii (1970) most ^{210}Po and ^{210}Pb in the layer of air near the earth is associated with dust particles. The ^{210}Pb and ^{210}Po content in this layer depends on the latitude, the altitude above sea level and the character of the surface (ocean, continent).

The largest contents of these radionuclides are found in the atmosphere at temperate latitudes. In equatorial regions it is smaller and is particularly low above the poles. Moreover in the Northern Hemisphere the content of ^{210}Po and ^{210}Pb is higher than in the south. Presumably, this distribution is due to the differences in the land and ocean areas. The principal source of radon, and consequently ^{210}Po and ^{210}Pb, are continental rocks.

The distribution of ^{210}Pb in soils is not the same as that of the parent ^{226}Ra. Usually the upper layer (0-5 cm) has a higher concentration of daughter radionuclides, which is presumably due to the supply as aerosols.

The ^{210}Po and ^{210}Pb enter the surface water with atmospheric precipitates and due to leaching from soil and rock. The concentration of ^{210}Pb in the upper 100 m layer of water of the ocean is 25 times greater than what can be formed from the ^{226}Ra contained therein. This makes us think that most of the ^{210}Pb comes to the ocean from the atmosphere. The time during which ^{210}Pb is removed from the upper layer of the ocean is 2-5 years. For this reason continental water in lakes and rivers contains substantially more ^{210}Pb than water in deep wells (underground water). As a rule, bottom deposits have several times more

^{210}Pb and ^{210}Po than the parent ^{226}Ra, which indicates that they come from the atmosphere. The ratio of $^{210}Po/^{210}Pb \leq 1$.

Underground water does not contain excess ^{210}Pb and ^{210}Po as compared to ^{226}Ra or ^{222}Rn. There is no marked difference between the ^{210}Pb or ^{210}Po concentrations and equilibrium ^{226}Ra in rock. Excess ^{210}Po was discovered in volcanic gas discharged during eruption.

^{210}Po and ^{210}Pb also enter the biosphere by aeral means. Over 80% of ^{210}Pb and ^{210}Po enter plants in the same way. The $^{210}Po/^{210}Pb$ ratio in the tissues of annual plants is 0.23, which corresponds to their ratio in the air. In perennial plants ^{210}Po accumulates from ^{210}Pb, and their ratio is close to unity. Owing to its volatility, ^{210}Po is sublimated in tobacco smoke, and when it enters the lungs of a smoker it is sorbed in them. Elevated concentrations of ^{210}Pb and ^{210}Po were discovered in lichens (Icelandic moss). Icelandic moss is the staple of reindeer, which in turn is the staple for the inhabitants of the North. The source of the radionuclides in the lichens is the atmosphere, since soil alimentation is insignificant. Lichens contain 10 times less ^{226}Ra than ^{210}Pb and ^{210}Po. The specific concentration of ^{210}Po and ^{210}Pb in organisms dwelling in water is higher than that in their habitats. The concentration factor for ^{210}Pb is of the order of 10^4, and somewhat higher for ^{210}Po. The content of radionuclides in zooplankton is higher than in phytoplankton [5.6].

In terrestrial animals ^{210}Pb and ^{210}Po is concentrated in bones; the concentrations in the muscles and other parts of the body are tens and hundreds of times lower. However, the $^{210}Po/^{210}Pb$ ratio in soft tissues increases, which shows the direct accumulation of Po in them. In bones ^{210}Po is formed from ^{210}Pb.

5.2.4 Protactinium

In nature two isotopes of protactinium have been found $^{231}_{91}Pa$: ($T_{1/2}$ = 3.43 × 10^4 years) from the ^{235}U series and the two isomers $^{234}_{91}Pa$ ($T_{1/2}$ = 6.7 hours and $T_{1/2}$ = 1.175 min) from the ^{238}U series.

Protactinium is an actinide and its chemistry is very close to Th^{4+} and U^{4+}. On the other hand, it has much in common with the elements in Group 5A, being a chemical analogue of tantalum. Some of its properties are close to those of Zr and Hf. Separation of Pa from the solutions by coprecipitation shows that Ta is a worse coprecipitant with Pa than Th and Zr.

The principal oxidation number of Pa is 5, which occurs in its highest oxide and halides. Pa isomorphously cocrystallizes with many Ta(V) compounds, it is more electrically positive than Ta and does not form compounds analogous to tantalates. However, Pa compounds with oxygen and halogens have been obtained, where it has an oxidation number of four, and three in the solution. In solution it is isomorphic with Th(IV) ($R_{Pa^{4+}}$ = 0.96 Å, $R_{Th^{4+}}$ = 0.99 Å). The Pa^{4+} ion is much more easily oxidized to Pa(V) than uranium U^{4+}. It oxidizes in air. Protactinium is a chemically inert element. Its compounds dissolve poorly and undergo hydrolysis. In strongly acidic solution Pa^{5+} is found in a hydrolyzed form. At very low concentrations of Pa in solution, which is typical in nature, the hydrolysis products are rapidly adsorbed on colloidal particles of other substances to form pseudocolloids, and on solid surfaces. Pa cations readily form complex compounds, though most of them are as unstable to hydrolysis as the simple compounds. In aqueous solutions only the fluoride, sulphate and some organic complex compounds are relatively stable.

In nature Pa is encountered in minerals of U(IV), being a product of the ^{238}U and ^{235}U decay. Only by the long-lived isotope ^{231}Pa is valuable in geochemistry. The average content of ^{231}Pa in the earth's crust, calculated from the uranium content, is 0.9×10^{-10} weight per cent. In endogenic minerals, Pa is found in radioactive equilibrium with uranium. When uranium is leached out from rocks by natural water we would expect the solid phase to be richer in ^{231}Pa with respect to the uranium. In young volcanic rocks ^{231}Pa is at smaller concentration with respect to ^{235}U [5.4], because of the different volatilities of these elements in volcanic gases. In natural water Pa is formed due to the decay of the dissolved uranium. Owing to intensive hydrolysis, it rapidly sorbs on suspended and colloidal particles, and precipitates with them on the floor of the water basin. This process is most marked in the ocean, where in the water there is a deficit of ^{231}Pa with respect to ^{231}U, while bottom deposits, conversely, contain more ^{231}Pa.

5.3 The Disturbance in the Radioactive Equilibrium in Decay Series

5.3.1 The Basic Principles of Radioactive Equilibrium

All the members of a radioactive series are linked by irreversible and mutual transformations. If a system containing radionuclides of a particular series is closed, i.e. there is no removal or supply of individual members, then the radioactive equilibrium sets in (see Ch. 1).

In general, the rate at which equilibrium establishes in a series is determined by the half-lives ($T_{1/2}$) of the most long-lived members. With an accuracy of

0.8%, equilibrium is established within time $7T_{1/2}$. Taking into account the analytical error in determining radionuclides, $7T_{1/2}$ will correspond to radioactive equilibrium. In the ^{238}U series the most long-lived intermediate member of the series is ^{234}U ($T_{1/2}$ = 2.48 × 10^5 years). Given above, the equilibrium in the ^{238}U series will occur after 2.48 × 10^5 years × 7 ≈ 1.7 mln years. In the same way, the equilibrium in the ^{235}U series will be established in 3.4 × 10^4 × 7 = 2.4 × 10^5 years (where 3.4 × 10^4 = $T_{1/2}$ ^{231}Pa). The ^{232}Th series will be in a state of equilibrium after 37 years ($T_{1/2}$ for the most long-lived member of the ^{228}Ra series is 5.75 years since 5.75 × 7 ≈ 37 years). Given a radioactive equilibrium in a decay series, the number of atoms of each member and its rate of decay are related as follows: $N_1\lambda_1 = N_2\lambda_2 = ... = N_i\lambda_i$ or

$$N_1 / T_1 = N_2 / T_2 = \ldots = N_i / T_i \quad (5.2)$$

5.3.2 Mechanisms of Radioactive Disequilibrium

Radioactive equilibrium in a decay series is considered to be disturbed if the relationship between the members of the series does not satisfy Eq. (5.2). The disturbance of equilibrium is due to migration of members of the series relative to one another from the system or into the system. The radionuclides must be moved from the system or into it by a distance exceeding the size of the system, and within a time shorter than the time required for radioactive equilibrium to be established.

Depending on how long the system was opened, and then closed, or on the rate at which it is opened, we can observe various relationships between individual members of the series. (a) If after the disturbance of the equilibrium the system remained closed within 10^7 or more years, then in all three natural series of decay radioactive equilibrium is established; (b) if the system closed 10^3-10^4 years ago, then we shall only observe the disequilibrium relationships between the parent of the series and the most long-lived daughter members of the series:

$$^{238}U\text{–}^{234}U\text{–}^{230}Th\text{–}^{226}Ra \text{ and } ^{235}U\text{–}^{231}Pa$$

(c) When we deal with processes that last less than 10 years, or with seasonal variations (months), the radionuclides are supplemented by ^{210}Pb, ^{210}Po (series ^{238}U); ^{227}Ac (series ^{235}U), and the chain ^{232}Th-^{228}Ra-^{228}Th (series ^{232}Th).

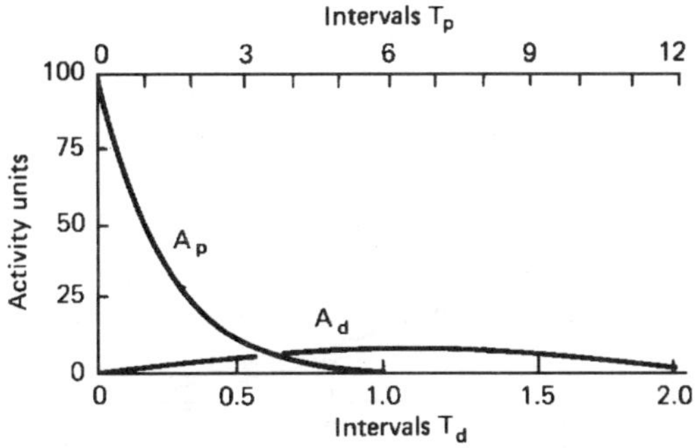

Fig. 5.1. Variation in the activity of short-lived parent A_p and long-lived daughter A_d of radionuclides in time

(d) Disequilibrium ratios between short-lived products at the end of the series can be found when geochemical processes violating equilibrium proceed continuously and rapidly. Thus we can assess the geochemical use of disequilibrium radioactive systems and the choice of radionuclides to suit the research.

Pairs of radionuclides that are genetically related are usually used, for instance, $^{234}U/^{238}U$, $^{226}Ra/^{238}U$, $^{230}Th/^{234}U$; and $^{228}Th/^{228}Ra$. The degree of disequilibrium is determined either from the activity ratio between daughter A_d and parent A_p atom (A_d/A_p), or from the following relation: $\delta d = \frac{A_d - A_p}{A_p}$ 100%. It follows from the definition that in the case of the radioactive equilibrium the indicated values will be $\frac{A_d}{A_p} = 1$ and $\delta d = 0$.

Depending on the ratio of the half-lives $T_{1/2}$ of the parent and daughter radionuclides, several cases may take place.

1. $T_{1/2}(p) < T_{1/2}(d)$: the parent radionuclide disintegrates more rapidly than the daughter one. In this case no equilibrium is possible if the parent radionuclide is isolated from a preceding member that is more long-lived in the series. The activity of the parent radionuclide will decrease as $N = N_0 e^{-\lambda t}$. The activity of the daughter radionuclide will rise to a maximum considerably less than that of the parent one. At a time well after most of the parent has disappeared, the daughter activity will decay at a rate of the daughter (Fig. 5.1). An example is the pair ^{234}Th (24 days)-^{234}U (2.5×10^5 years).

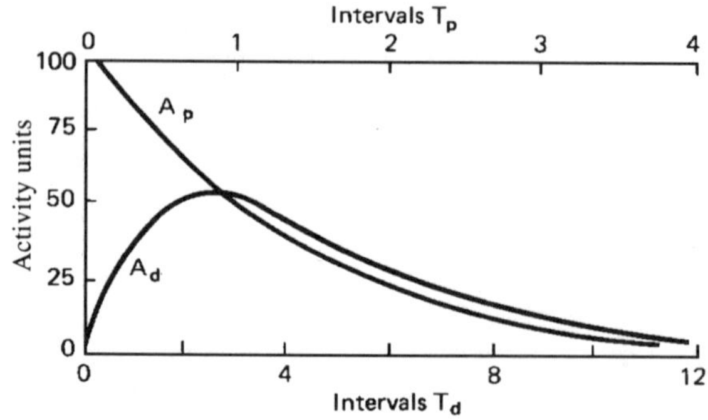

Fig. 5.2. Variation in the activity of parent A_p and daughter A_d of radionuclides when $T_p > T_d$

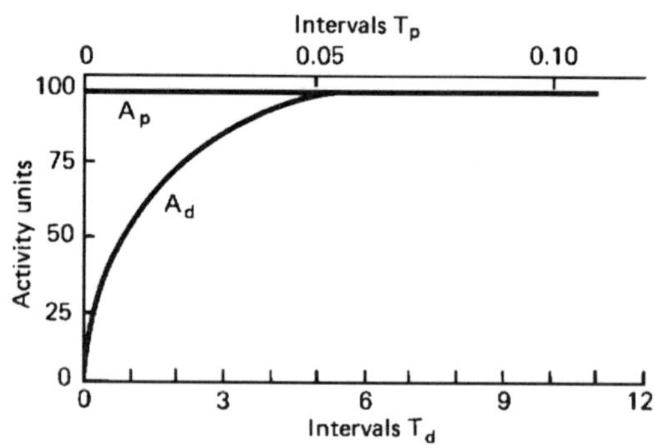

Fig. 5.3. Variation in the activity of A_p and daughter A_d of radionuclides in the case when $T_p >> T_d$

2. $T_{1/2}(p) \geq T_{1/2}(d)$. If the half-life of the parent radionuclide is commensurable with the half-life of the daughter radionuclide or slightly greater, the daughter activity increases a level of transient equilibrium with its parent. Then two activity curves decrease together, at a rate determined by the decay rate of the parent (Fig. 5.2). An example is the pair ^{228}Ra (5.75 years)-^{228}Th (1.9 years).

3. $T_{1/2}(\text{p}) >> T_{1/2}(\text{d})$. Most often we use radionuclide pairs in which the half life of the parent radionuclide is much greater than that of the daughter radionuclide, e.g., ^{226}Ra (1.6×10^3 years)-^{222}Rn (3.8 days). In this case and within the time determined by $T_{1/2}(\text{d})$, a 'secular equilibrium' becomes established and the relationship of the activities of the nuclides will be determined by the equation: $A_d/A_p = 1 - e^{-\lambda_d t}$. It is clear from the relationship $N_p\lambda_p = N_d\lambda_d$ that the greater the rate of the decay of the daughter radionuclide, the smaller is the number of its atoms in equilibrium with the parent radionuclide (Fig. 5.3).

4. We rarely find in natural systems the cases described above when the daughter radionuclide is absent ($A_d = 0$) at time ($t = 0$). Some amount of it usually occurs together with the parent atoms, although $A_d \neq A_p$. A change in the half-live of a daughter nuclide will be expressed in the following equation:

$$A_\text{d} = A_\text{d}(0)e^{-\lambda_\text{d}t} + A_\text{p}\left(1 - e^{-\lambda_\text{d}t}\right) \tag{5.3}$$

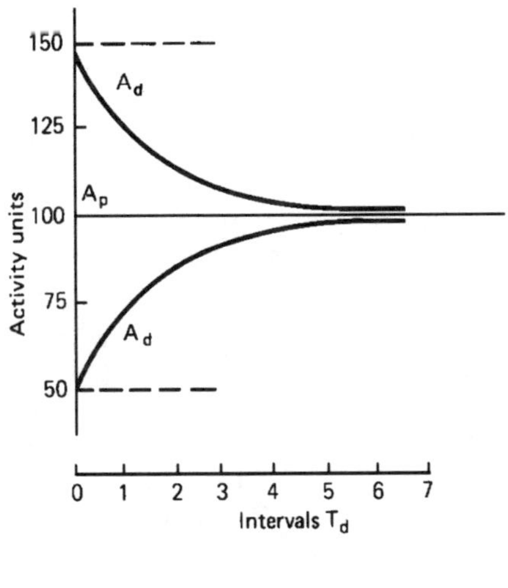

Fig. 5.4. Two variants of disequilibrium systems with $T_p \neq T_d$

The first component corresponds to the decay of the daughter atoms, while the second component expresses their accumulation from the parent ones. There are two cases here: $A_d > A_p$ and $A_d < A_p$, which can be clearly seen in Fig. 5.4.

5.3.3 Factors Causing Disturbance in Radioactive Equilibrium

The disturbance in the radioactive equilibrium arise in an open system when the mobilities of the daughter and parent atoms are not the same in the system's geochemical environment. In most cases, the radioactive equilibrium is disturbed at an interaction, e.g., at the boundary between two systems. In nature such interactions are frequently created during contact between water in pores and fissures and rocks, friable sediments or soils. Depending on the composition of the ions dissolved in water, the pH and Eh, the daughter nuclides may be soluble or not. As a result of selective leaching radionuclides in the same decay series will be differentiated. The liquid phase will be richer in the most soluble members of the series. On the contrary, the solid phase will be made poorer. The result is that rock and the water circulating through it will display a disturbance of the radioactive equilibrium in the decay series. The selective sedimentation of radionuclides from natural water and the subsequent adsorption on the solid phase are also factors that disturb radioactive equilibrium between the members of a single series.

There are two main reasons for differentiation between the members of a series: (1) different chemical properties of radionuclides and/or (2) different forms of existence.

I. The chemistry of radionuclides in natural decay series was discussed in Chapter 3 and elsewhere above. These data indicate how certain radionuclides behave in particular geochemical conditions. Several examples are adduced below.

(1) In weathered rocks, natural waters will oxidize the uranium atoms, which lie on the surface of mineral grains and transfer them into solution. In addition the daughter atoms of the thorium isotopes (^{234}Th and ^{230}Th) have much lower solubilities and will remain in their place. This means that rocks being weathered, and particularly soils, acquire disequilibrium relationships of activities with excess ^{230}Th (^{230}Th/^{234}U ≥ 2.0), while waters, conversely, have relatively less ^{230}Th (^{230}Th/^{234}U ≤ 0.5).

(2) ^{238}U and ^{234}U in oceanic water are in the solution as stable carbonate complexes, and the daughter radionuclide ^{230}Th, which is formed from ^{234}U, will adsorb on the suspended particles or coprecipitates with iron hydroxides,

thus enriching the bottom deposits. This means that the radioactive equilibrium will be disturbed both in water and in the sediments.

(3) An example of the another geochemical situation is the interaction between underground waters and water-enclosing rocks in hydrocarbon deposits, which create strongly reducing conditions. Uranium is not oxidized in this case and so does not dissolve in water. At the same time, the daughter radionuclide ^{226}Ra is readily leached from rocks and is stable in brines not containing barium sulphate. As a result, the radioactive equilibrium at the rock-water interface is considerably disturbed, and the activity ratio of $^{226}Ra/^{238}U$ in the brines can reach 10^2-10^4.

II. The different mobilities of the daughter and parent atoms become particularly apparent for pairs of isotopes of the same element, which do not have different chemical properties, e.g. $^{234}U/^{238}U$, $^{228}Th/^{232}Th$, $^{224}Ra/^{228}Ra$. During leaching of the rocks or their weathering products, the daughter isotopes more easily migrate into solution than their parents. In the active water-exchange zone, the activity ratios of the isotopic pairs typical of water are usually greater than the equilibrium value (unit), whereas for water-inclosing rocks, they are less than unit. The difference between the isotope mobilities for the same element in a rock-water system is that the daughter products form as a result of radioactive decay, while the parent products enter the mineral during crystallization. The principal factor affecting mobility is the alpha recoil the daughter atom experiences. In accordance with the momentum conservation, the daughter nucleus resulting from alpha-decay acquires a radioactive recoil energy E_{rec}, which is as many times smaller than the energy of alpha particle E_α as the mass M_{rec} is greater than the mass of the alpha particle M_α:

$$E_{rec} = \frac{E_\alpha \times M_\alpha}{M_{rec}}$$

For instance, in ^{232}Th decay, which has the smallest energy of the alpha particles of all the members of the three series, the nucleus ^{228}Ra forms. It has a recoil energy $E_{rec} = \dfrac{4.0 \text{ MeV} \times 4}{228} = 70$ KeV, which is sufficient to cover the distance of 200 Å in the crystalline lattice of the mineral [5.5]. For a radioactive series the overall path a radionuclide will travel is determined by the number of preceding alpha decays. This displacement of atoms is a 'recoil diffusion' and is the function of

(a) the number of α-decays in the chain;
(b) the energy of individual decays;
(c) the gradient of concentrations of the parent radionuclides.

The mobility of daughter radionuclides is promoted by the fact that when recoil atoms in the lattice of a mineral leave the structural position occupied by the parent atoms, they become impurity atoms. At the edge of a mineral grain or in microfissures they are more easily leached. Moreover, the deformed lattice structure is favourable for penetration by water along the track of the recoil atom and the development of secondary processes.

The enhanced mobility of daughter radionuclides as a result of α recoil is proved by the transition of ^{234}Th from zircons into water. Note that the zircon grains do not dissolve in the process (Kigoshi).

In some cases the disturbance of radioactive equilibrium occurs between two solid phases. For instance, the diffusion of radon from zircon grains into the enclosing rock that is promoted by the radioactive recoil will lead to the accumulation of radiogenic lead outside the system, i.e. outside the zircon crystal. This results in the rejuvenation of the zircon, which emerges due to the recoil diffusion of radon rather than as a consequence of the thermal diffusion of lead atoms.

5.4 Disequilibrium Isotopic Methods of Dating

5.4.1 Ionium and Ionium-Protactinium Methods

The ionium dating method uses the disturbance of the radioactive equilibrium between ^{234}U and its decay product ^{230}Th or ionium (Io). ^{230}Th has a half-life of 75 000 years, which determines the upper limit of the method is about 400 000 years. The lower limit is determined by the analytical error and is usually about 10 000 years. An interval of 10^3-5×10^5 years lies between the maximum possible age established by ^{14}C dating (about 40 000 years) and the minimum potassium-argon age (≈ 0.5 mln years). This reveals the importance of the ionium method in radiogeochronology.

Equation (5.1) is used to estimate the age by the ionium method:

$$^{230}\mathrm{Th}_t = {}^{230}\mathrm{Th}_0 e^{-\lambda_{230}t} + {}^{234}\mathrm{U}\left(1 - e^{-\lambda_{230}t}\right)$$

(^{230}Th and ^{234}U are expressed in activity units).

The equation contains two unknowns, viz. time (t) and initial concentration $(^{230}Th)_0$. The equation can be used with a view to the above conditions. The simplest systems are those in which at the initial moment of formation (t = 0) both ^{230}Th and ^{234}U were absent. In the first case the first term of the equation equals 0; in the second one it is 2. The latter case occurs but rare. We discuss some examples illustrating these systems:

I. $^{230}Th_0 = 0$

(a) Examples of a system in which at the beginning there was only uranium and practically no ^{230}Th are marine biogenic carbonates (corals and shells). To make their skeletons or shells, the organisms extract carbonate carbon from sea water, which is accompanied by the destruction of carbonate complexes containing uranium and the coprecipitation of uranium and calcium carbonate. The mechanism supplying uranium into biogenic carbonates is proved by the constant value of the ratio U/Ca in sea water and corals, which is nearly 2×10^{-6} [5.8]. Since at the initial moment of formation of a biogenic carbonate ^{230}Th is absent, Eq. (5.1) is simplified

$$\left({}^{230}\text{Th}/{}^{234}\text{U}\right)_t = 1 - e^{-\lambda_{230}t}; \text{ or } t = \frac{1}{\lambda_{230}} \ln\left[1 - \left({}^{230}\text{Th}/{}^{234}\text{U}\right)_t\right]$$

The 'purest' objects without any initial ^{230}Th are the biogenic carbonates of coral reefs. An analysis of coral reef age is used for paleoclimate reconstructions, often together with an analysis of the isotopic composition of oxygen. Sea shells buried in marine sediments can contain secondary radionuclides as impurities, which were supplied at diagenesis stage.

To avoid secondary pollution the shell is stripped layer-by-layer and each fraction is analyzed separately [5.9]. If the fractions corresponding to the outer and inner parts of the shell coincide in terms of their radionuclide ratios, while the ^{232}Th is absent, no terrigenous contamination has occurred. The analysis data of interior zones coincide more often, i.e. these zones are taken for the primary ones, i.e. those that are not contaminated. Mollusk shells help to identify the age of sea terraces, which is essential for paleogeographic reconstructions.

In some cases phosphatic nodules are a good object of study [5.10].

(b) Ionium dating of continental carbonates is very difficult. Continental systems are more heterogeneous and there is a great possibility of contamination by the radionuclides from ground water. Moreover, oceanic water has a constant $^{234}U/^{238}U$ ratio of 1.15, i.e. there does not appear any problem in establishing the original ratio. The variations in $^{234}U/^{238}U$ in ground water can be considerable, while the initial ratio will affect the rate of ^{230}Th accumulation in samples that are more than tens of thousands years old. Therefore, the calculation formula has to take into account the decay of ^{234}U:

$$\left({}^{230}\mathrm{Th}/{}^{234}\mathrm{U}\right)_t = \frac{1-e^{-\lambda_{230}t}}{\left({}^{234}\mathrm{U}/{}^{238}\mathrm{U}\right)_t} + \frac{\lambda_{230}}{\lambda_{230}-\lambda_{234}} \times \left(1-\frac{1}{\left({}^{234}\mathrm{U}/{}^{238}\mathrm{U}\right)_t}\right) \times \left(1-e^{-(\lambda_{230}-\lambda_{234})t}\right)$$

Good samples are for dating speleolites containing less than 1% of an insoluble residue and practically devoid of the primary isotopes of thorium [5.11]. The dating of speleolites (e.g. stalactites) makes it possible to determine the age of caves and to establish the chronological sequence of paleoclimatic variations in the past. For instance, the growth of speleolites in preglacial and glacial stage was retarded or ceased due to the absence on the surface of liquid water that could penetrate the caves, dissolve the limestone and precipitate $CaCO_3$. The period within which speleolites stop growing corresponds to glacial periods. We can obtain reliable data from the $^{230}Th/^{234}U$ ratio for mineralized bone remains buried in river terraces [5.2]. However, the rapid secondary migration of uranium requires rigid conditions for the system to remain closed. These conditions are only fulfilled in permafrost rocks [5.13].

II. $(^{230}Th)_0 >> (^{234}U)_0$

Enrichment of ^{230}Th with unsupported uranium can be found in oceanic sediments. Similar enrichment occurs in ferromanganese nodules. The coprecipitation of ^{230}Th formed in sea water with Fe and Mn hydroxides enriches the nodules with unsupported ^{230}Th, whose nodules diminish from the periphery towards the centre of nodules due to decay. To date a nodule (its central part), i.e. the time it began to grow, the primary factor is the current ratio of $^{232}Th/^{234}U$, for which the value on the outer surface layer of the nodule is taken. When we determine the rate at which the nodule developed we must assume that the ratio of nuclides precipitated from water is constant during the time it grows. Establishing the age and rate of sedimentation of oceanic sediments is much more complicated, since it requires a whole range of assumptions [5.12].

1. The content of ^{230}Th and ^{234}U in the ocean is constant within the last half-million years, i.e. within the interval of time determined by the ionium method.

2. The rate at which ^{230}Th precipitates on the ocean floor is invariable.

3. There is no migration of ^{230}Th and ^{234}U in the sediments.

4. The distribution of sediments in the column should not be disturbed as regards the initial arrangement of layers.

It is difficult to determine how valid these assumptions are in each particular case. The truest results were obtained from deep sediments distinguished by a low rate of sedimentation ranging from 0.1 to 10 g/cm^2 over 10^3 years [5.7]. We cannot say with any degree of accuracy what the initial ratio of the radionuclides was.

The calculation implies the excess of activity of ^{230}Th over that of ^{234}U, i.e. $\Delta^{230}\text{Th} = {}^{230}\text{Th} - {}^{234}\text{U}$; $\Delta^{230}\text{Th}_i = \Delta^{230}\text{Th}_0 e^{-\lambda_{230}t}$, where $\Delta^{230}\text{Th}_0$ is the excess of ^{230}Th activity on the surface of the ocean floor and $\Delta^{230}\text{Th}_i$ is that for the i-th horizon. If the change in ^{230}Th with the depth of the column corresponds to the radioactive decay curve, it is possible to approximate the average rate of sedimentation. To do this it is necessary to find the depth of sediments where $\Delta^{230}\text{Th} = 0$ and divide it by 450 thousand years [5.14].

Ionium-Protactinium Method

The ^{230}Th/^{234}U ratio is largely dependent on the secondary migration of uranium. A dating method that is more reliable than ionium-uranium dating is ionium-protactinium dating. ^{231}Pa is formed by the decay of ^{235}U associated with ^{238}U, which gives rise to ^{230}Th, by a constant ratio. Given the chemical inertness of protactinium, we can assert that the mechanism by which ^{231}Pa enters deep sea sediments from the water, where it is formed, will be similar to the coprecipitation of ^{230}Th and due to rapid hydrolysis [5.14]. ^{230}Th and ^{231}Pa sorbed on a terrigenous suspension or contained in colloidal aggregates of the Fe and Mn hydroxides fall together to the ocean floor. The behaviours of these radionuclides in the bulk of the sediment are similar. The distribution curves for their ratios correlate more with the curves of radioactive decay, than with the distribution curves for D^{230}Th or D^{231}Pa along the column, which are affected by the migration of uranium. This proves that (Fig. 5.5) during sediment diagenesis there is no noticeable separation of these radionuclides. The age is calculated from equation

$$t = \frac{2.303}{\lambda_{231} - \lambda_{230}} \times \log\left[\frac{\left({}^{230}\text{Th}/{}^{231}\text{Pa}\right)_t}{\left({}^{230}\text{Th}/{}^{231}\text{Pa}\right)_0}\right]$$

where λ_{231}, λ_{230} are the decay constants; $({}^{230}\text{Th}/{}^{231}\text{Pa})_t$ is the ratio of the radionuclides in the sample; $({}^{230}\text{Th}/{}^{231}\text{Pa})_0$ is the ratio of the radionuclides in sea water. Owing to the difference in the half-lives, the radionuclide ratio will change.

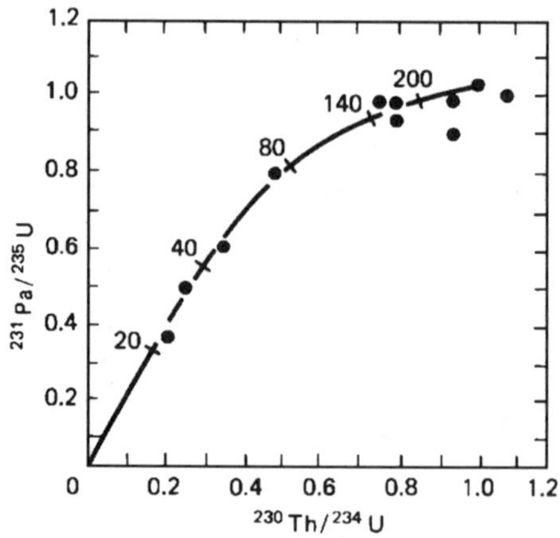

Fig. 5.5. The relation of $^{230}Th/^{234}U$ as regards $^{231}Pa/^{235}U$ in phosphate nodules of the continental slope of Eastern Australia.

The curve draws an enclosed system calculated for the primary relations $^{230}Th/^{234}U = 0$; $^{231}Pa/^{235}U = 0$;$^{234}U/^{238}U = 1.14$. The figures above the curve indicate time in thousands of years (according to O'Brein et al. 1986)

Ionium Method for Dating Quaternary Volcanic Rocks

There is a disequilibrium between ^{230}Th and ^{234}U in recent volcanic rocks [5.15]. This disequilibrium has the endogenic cause although the mechanism of separation of uranium and thorium has not been fully studied. It is thought that the perturbation in the equilibrium occurs in volcanic eruption during the degasification of the melt due to different stabilities of volatile compounds of uranium and thorium [5.16]. Irrespective of the character of the radioactive equilibrium perturbation mechanism, the difference can be used to establish the age of quaternary volcanic rocks. It has been established that there is no division of uranium isotopes and the activity ratio is $^{234}U/^{238}U = 1$ [5.16]. Thus the age can be estimated from equation (5.1) without considering ^{234}U decay, i.e.

$$t = -\frac{1}{\lambda_{230}} \ln \left[\frac{\left(^{230}Th/^{238}U \right)_t - 1}{\left(^{230}Th/^{238}U \right)_0 - 1} \right]$$

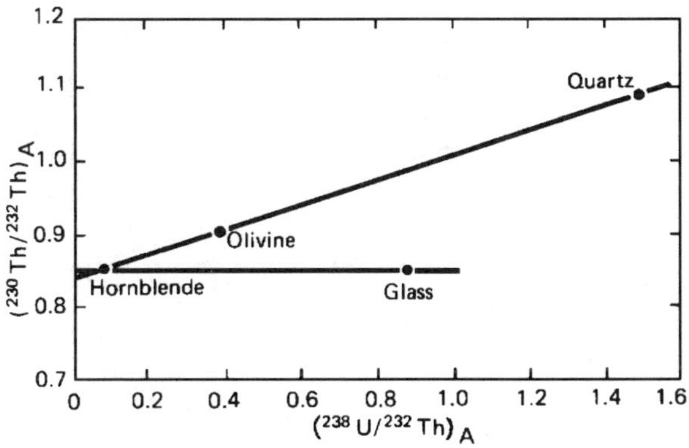

Fig. 5.6. Ionium-uranium isochrone of rhyolites of Mono Crater, California (according to C. Allègre, 1968)

Besides the age, the $(^{230}Th/^{238}U)_0$ ratio in the equation is unknown. It can be established by several methods. Cerrai et al. (1965) suggested that two different syngenetic minerals (or mineral and glass) should be analysed, and that two equations in two unknowns ($^{230}Th_0$ and t) should be solved. However, experiments demonstrate that minerals, and zircons in particular, are often coeval with the volcanic rock [5.15].

A more reliable method is to construct an isochrone in the coordinates of the ratios of the activities $^{230}Th/^{232}Th$ and $^{238}U/^{232}Th$ by analyzing a number of minerals with different thorium-uranium ratio $(^{232}Th/^{238}U)$ (Fig. 5.6). Coetaneous minerals should have the same values of the initial $(^{230}Th/^{232}Th)_0$ ratio, which corresponds to the point where the isochrone intersects ordinate axis. The slope tangent will be a function of age: $\tan \alpha = 1 - e^{-\lambda_{230}t}$. It is possible, in principle, to take the average value of $^{230}Th/^{234}U$ in rocks of contemporary (historical) eruptions of the same volcano, which are similar in composition to the analyzed one and have the same Th/U ratios, as $(^{230}Th/^{234}U)_0$ [5.17].

5.4.2 Uranium-Isotopic Dating

The uranium-isotopic method uses the ratio of genetically bound uranium isotopes $^{234}U/^{238}U$ in the same decay series. It covers a time interval of about 1 mln years. The method has been used to date marine biogenic carbonates and has much in common with the ionium method. The advantage

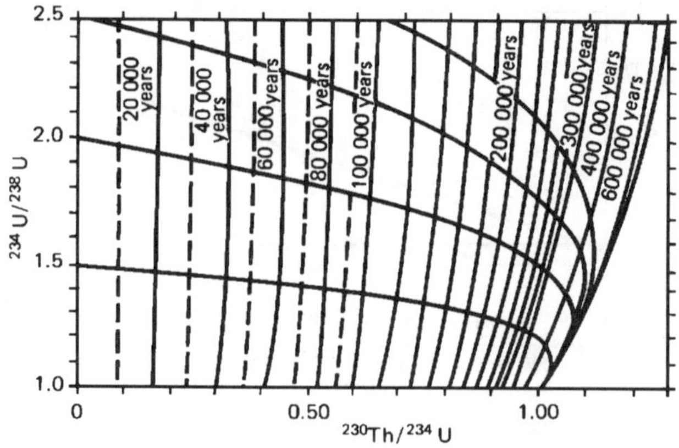

Fig. 5.7. Variations of $^{234}U/^{238}U$ and $^{230}Th/^{234}U$ relations of activities in time for the enclosed system, free of original ^{230}Th.

Near-vertical lines are the isochors for the samples of the same age, but different initial relation of $^{234}U/^{238}U$. Horizontal curves show the age of parts of samples with ratios $^{234}U/^{238}U$ = 1.5, 2.0, 2.5 (according to Osmond, Cowart, 1982)

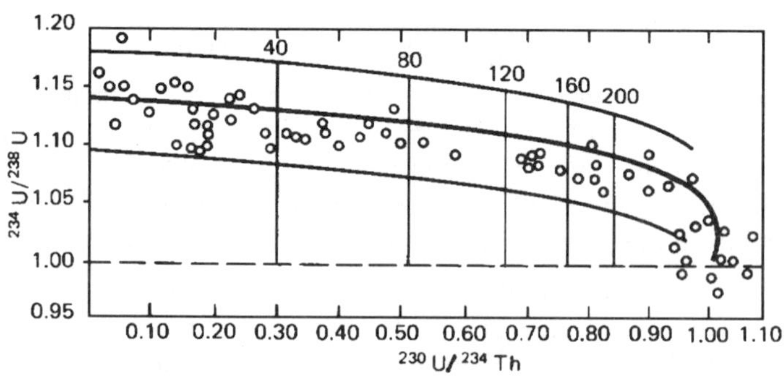

Fig. 5.8. Relation of $^{230}Th/^{234}U–^{234}U/^{238}U$ activities in phosphorites of the continental slope of Eastern Australia. The congruity of experimental points with the theoretical curve indicates the closeness of the system in time, calculated for the initial relations of $^{230}Th/^{234}U = 0$ and $^{234}U/^{238}U = 1.14$.

Figures above the curve correspond to age in thousands of years. Thin lines are correlated with $\pm 2\sigma$ ($1\sigma^{234}U/^{238}U = \pm 0.02$) (according to O'Brien et al. 1986)

of the method is that it is precisely known what the initial ratio between isotopes in oceanic water is, viz., $\gamma_0 = (^{234}U/^{238}U)_0 = 1.15$. Equation (5.1) yields a formula for the age [5.15]

$$t = \frac{1}{\lambda_{234}} \times \ln\left(\frac{\gamma_0 - 1}{\gamma_t - 1}\right)$$

The $\gamma_t = (^{234}U/^{238}U)_t$ is the isotope ratio in the sample at the time of measurement, γ_0 is the ratio at the initial moment of time and λ_{234} is the ^{234}U decay constant. For continental systems, the method has not been applied because there are no large reservoirs on land comparable with the ocean where the isotopic ratio of uranium is constant for such a long time. This is shown in Figs. 5.7 and 5.8 by an example illustrating the joint use of the ratios $^{234}U/^{238}U$ and $^{230}Th/^{234}U$ in determining the age of phosphate nodules and the evidence of the closure of the system.

5.4.3 ^{210}Pb Method of Dating

Dating using Pb^{210} ($T_{1/2}$ = 22 years) is employed for relatively young geological formations, viz. about 100 years old (glaciers, lacustrine and fluvial deposits, etc.). The principle underlying the method is that Pb^{210} unsupported by the parent radionuclei precipitates on the earth's surface from the atmosphere together with aerosols or is washed out from sediments [5.18]. ^{210}Pb is constantly formed in the atmosphere from ^{222}Rn, which enters the atmosphere as a result of emanation from soils of rocks. ^{210}Pb decay follows the common law for a single radionuclide. The age of a sediment is determined from

$$t = \frac{2.303}{\lambda_{210}} \times \log\left(\frac{^{210}Pb_0}{^{210}Pb_t}\right)_A$$

where $^{210}Pb_0$ is the activity of lead on the surface of the Earth (h = 0), and $^{210}Pb_t$ is the activity in the sample at a depth h. The ^{210}Pb concentration is usually determined from its decay products, viz. β is the activity of ^{210}Bi ($T_{1/2}$ = 5.5 days), or α is the activity of ^{210}Po ($T_{1/2}$ = 138 days).

5.4.4 Dating Underground Water By Disequilibrium Isotopic Methods

The 'age' of underground water is the time passed after it infiltrated the earth's surface. When dating underground water it is essential that we should observe the same conditions as when dating solid samples:

(1) the age of the water should be of the same order as the half-life of the daughter radionuclide being used;

(2) the initial ratio of the radionuclides in the infiltrating waters should be known;

(3) the change in the radionuclide ratio should only be affected by radioactive decay and not be influenced by physical or chemical action.

The main difficulty in dating underground water by nonequilibrium isotopic methods is to observe conditions (2) and (3). The initial isotopic ratio is found by two ways. More often this quantity is determined as the isotopic ratio for the water catchment area. The other way is the approximation of the decay curve using several points given that condition 3 is observed. The principal problem is to prove the closure of the system. To date underground water we use the radionuclide ratios $^{222}Rn/^{226}Ra$, $^{234}U/^{238}U$, $^{228}Ra/^{226}Ra$, etc. In all these cases the influence of isotopic exchange in the rock-water system, the effect of leaching, sorption during changes in the physical chemistry or geochemical changes in the medium along the path of the underground water cannot be excluded. Therefore, the ages that are obtained are tentative. A more reliable age estimate for the movement of underground water comes from cosmogenic radionuclides (Ch. 7).

5.4.5 Determining the Rate of the Oil Well Plugging

Oil pools are often exploited by displacing oil with water whose composition sometimes differs from that of the oil field brine. The mixing of water with a different composition often precipitates poorly dissolved salts, $CaCO_3$; $Ca(Ba, Sr)CO_3$; $Ca(Ba, Sr)SO_4$ on the inner surfaces of the well, which in the long run leads to its plugging.

Isotopes of radium with Ba and Ca are coprecipitated. The concentration of the radium in the water at the water-oil interface is thus much larger. Thus, the formation of salt in the well is accompanied by an increase in radioactivity (the radiogeochemical effect). This effect can be used to discover from repeated gamma-ray log diagrams where degradation is occurring. The rate at which the well degrades can be determined by the disequilibrium isotopic method on the radionuclide ratio of $^{228}Th/^{228}Ra$:

$$t=\frac{1}{\lambda_{Ra}-\lambda_{Th}}\times\left[\ln\left(\frac{^{228}Th_t}{^{228}Ra_t}-\frac{\lambda_{Th}}{\lambda_{Th}-\lambda_{Ra}}\right)-\ln\left(\frac{^{228}Th_0}{^{228}Ra_0}-\frac{\lambda_{Th}}{\lambda_{Th}-\lambda_{Ra}}\right)\right]$$

where $^{228}Th_t$ and $^{228}Ra_t$ are the activities of radionuclides in the youngest (inner) parts of the 'well stone', $^{228}Th_0$ and $^{228}Ra_0$ are the activities of radionuclides at the initial moment of salt formation (at the well walls) and λ_{Ra} and λ_{Th} are the decay constants of ^{228}Ra and ^{228}Th, respectively.

The formula was derived from (1.25).

5.5 The Systematics of Uranium and Thorium Isotopic Ratios

5.5.1 Systematics of $^{234}U/^{238}U$ and $^{228}Th/^{232}Th$ Isotopic Ratios at the Lithosphere-Hydrosphere Interface

We mentioned in Sec. 5.3 that the differentiation of genetically bound isotopes of uranium ($^{234}U/^{238}U$) and thorium ($^{228}Th/^{232}Th$) occurs at the interface of the solid and liquid phases during their interaction, i.e. at the interface between the lithosphere and the hydrosphere. As a result of the differentiation of isotopes of radioactive elements the liquid phase gets richer in the daughter isotopes (^{234}U and ^{228}Th) while the solid phase gets poorer. In the end natural water will have activity ratios of $^{234}U/^{238}U > 1$ and $^{228}Th/^{232}Th > 1$. The rock or the products of its destruction where phases interact typically have opposite deviations from radioactive equilibrium, i.e., $^{234}U/^{238}U < 1$ and $^{228}Th/^{232}Th < 1$. The classification of these values is given in the histograms of Fig. 5.9. The patterns are substantiated by the experiments on the leaching of rocks by a weak solution of sodium chloride, which does not destroy the mineral lattice. In addition the leaching product is always richer with the daughter isotopes.

The precipitation of uranium and thorium from the solution leads to the enrichment of the solid phase by the elements with a different isotopic composition corresponding to the solution. This means that bottom deposits, soils and other formations which have a polygenic nature contain different uranium and thorium fractions differing in isotopic composition. The mineral and fragmental fractions will be close to the products of the destruction of rocks ($^{234}U/^{238}U < 1$ and $^{228}Th/^{232}Th < 1$), while the sorbed fraction will approximate the isotopic composition of water ($^{234}U/^{238}U > 1$ and

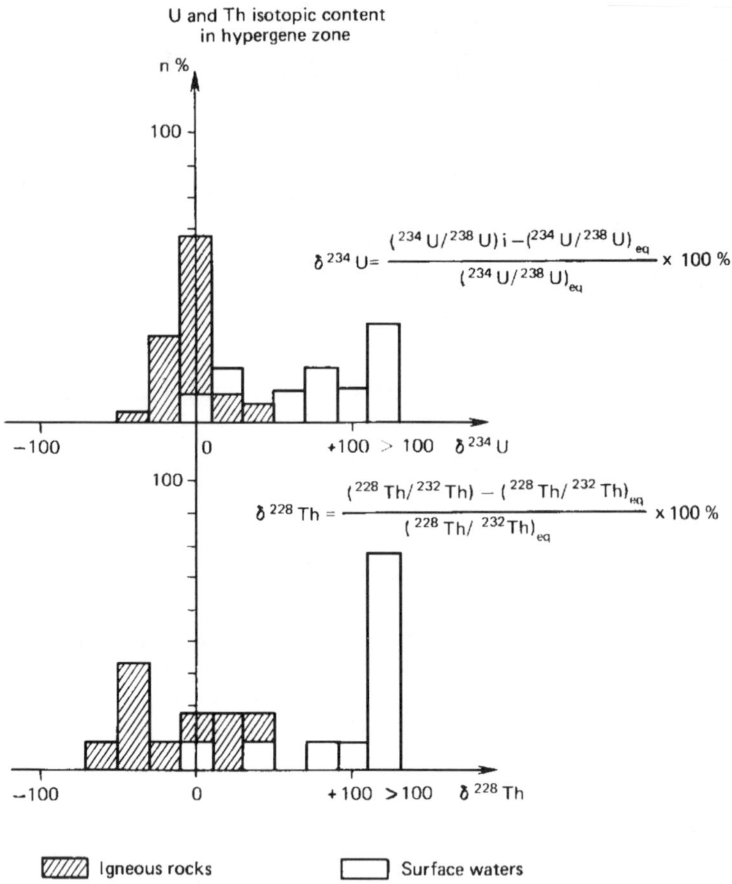

Fig. 5.9. Distribution histograms of isotopic relations of $^{234}U/^{238}U$ and $^{228}Th/^{232}Th$ in the system rock-water:

1—igneous rocks; *2*—surface waters

$^{228}Th/^{232}Th > 1$). The overall uranium and thorium in various fractions gives isotopic compositions close to the radioactive equilibrium ($^{234}U/^{238}U = 1$ and $^{228}Th/^{232}Th = 1$). The radioactive equilibrium in isotopic pairs in cases of this kind, which is particularly common in soils, does not indicate the absence of elemental migration, but is the result of mixing the fractions of uranium and thorium with different isotopic composition (Fig. 5.10).

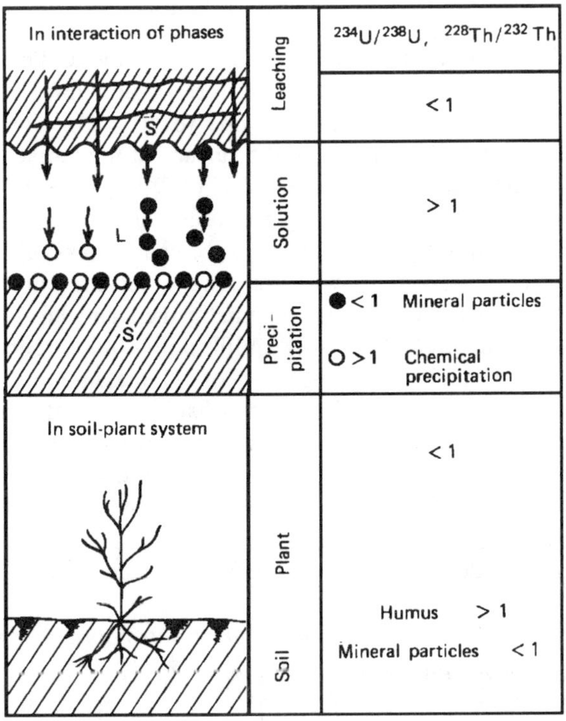

Fig. 5.10. Diagram showing the division of genetically connected isotopes of uranium and thorium in the zone of hypergenesis:
L—liquid phase; S—solid phase

The uranium and thorium isotopic systematics makes it possible to use the $^{234}U/^{238}U$ and $^{228}Th/^{232}Th$ ratios as geochemical indicators of elemental migration. It follows that if uranium and thorium in the solid phase (e.g. rocks, minerals, bottom deposits, and soils) have less daughter isotopes ($^{234}U/^{238}U < 1$ or $^{228}Th/^{232}Th < 1$), then leaching and removal of the element will be dominant. When we have excess daughter isotopes ($^{234}U/^{238}U > 1$ or $^{228}Th/^{232}Th > 1$) in the solid phase, most of the element is supplied as a result of precipitation from the solution.

Now let us consider uranium and thorium ratios used as indicators.

1. *Establishing the Closure of Geochemical Systems When Radiological Age Is to Be Determined*

(a) If an uranium-bearing mineral to be dated by the U-Pb method has recently (the past million years) suffered the loss of uranium (leaching), the $^{234}U/^{238}U$ activity ratio will be smaller than one ($\delta^{234}U < 0$). Conversely, the supply of uranium and its precipitation on the surface of microcracks and along the periphery of grains will produce the value of $^{234}U/^{238}U > 1$. Samples of this kind are unsuitable when establishing a uranium-lead age.

(b) When radiologically dating Quaternary volcanic rocks by the ionium method, the $^{234}U/^{238}U$ activity ratio will not always serve as an indicator of the closure of the system. The original magmatic melt and the rock that has formed from it are characterized by the equilibrium isotopic composition of uranium ($^{234}U/^{238}U = 1$) and similar position of all the uranium isotopes in structure of minerals [5.18]. Only after 40 thousand years will volcanic rock contain sufficient recoil ^{234}U atoms to produce a noticeable isotopic shift as a result of the leaching of uranium. Thus, for younger rocks the $^{234}U/^{238}U$ ratio cannot be an indicator of uranium leaching. It should be said that the intensity of uranium leaching from contemporary volcanic rocks is infinitesimal and is between 0.2-2.0%, i.e. does not exceed analytical error.

2. *Determining the Source of Uranium in Contemporary Hydrothermal Waters and Volcanic Exhalation*

In contemporary volcanic regions most hydrothermal sources have a disequilibrium ratio of the isotopes of uranium $^{234}U/^{238}U \neq 1$ [5.16, 5.17]. This indicates that the source of most uranium is more ancient volcanic rocks, from which the hydrothermal solutions have leached it. For minerals formed where hot volcanic gas emission (sublimate), the equilibrium isotopic composition of uranium has been discovered. For instance, $^{234}U/^{238}U = 1$ was characteristic of the craters of sublimates from the Great Fissured Tolbachic eruption in 1975-76 in Kamchatka. This means that the uranium had been removed from the boiling magma with the high-temperature gas and it is thus juvenile [5.24].

3. *Research of the Underground Water of the Deep Fractures in Alpine Folding Regions*

This region has a high seismic activity. During a shock, seismic and acoustic waves are formed. They open pores and porous waters become gravitational waters. This results in underground water with an abnormally high value of $^{234}U/^{238}U$, which reaches 8-12 with a background value of 1.2-2.5.

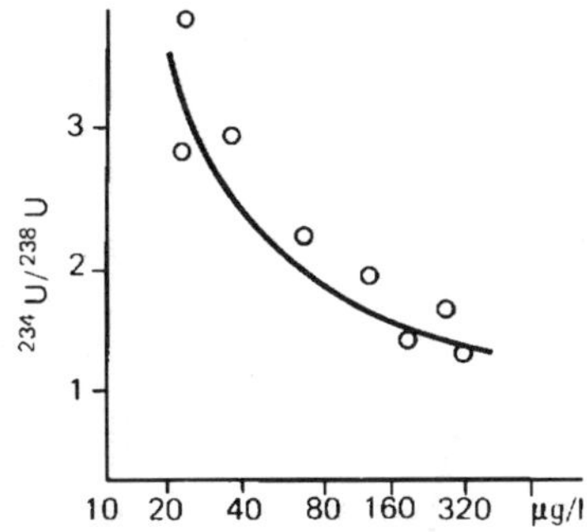

Fig. 5.11. Dependence of isotopic relation of uranium on its concentration in the waters of uranium-sulphide hydrothermal deposit (according to N.G. Syromyatnikov et al. 1976)

This phenomenon was established by research on the isotopic composition of uranium during the Tashkent earthquake, and by experimental analyses [5.20].

4. *How to Determine the Time Uranium Is in the Ocean*

The $^{234}U/^{238}U$ isotopic ratio in the ocean is constant and equal to 1.15 (1.14). If we accept the assumption that all U dissolved in the ocean is brought by the rivers and the average isotopic composition of uranium in the waters is 1.25, it becomes possible to calculate the approximate time uranium stays in the ocean [5.7]:

$$T = (1.25\text{-}1.15) : (1.15\text{-}1.00) \times \lambda_{234} = 250\ 000 \text{ years}$$

5. *The Prospecting of Uranium Deposits*

Where underground water interacts with uranium ores containing uranium minerals, water haloes of dispersion with high concentrations of uranium ($\geq 10^{-5}$-10^{-4} g/l) and near-equilibrium isotopic relation of $^{234}U/^{238}U \approx 1$ have been found to form [5.21]. An example is given in Fig. 5.11. The reason for this phenomenon is that uranium oxides (pitchblendes, uraninites, and blacks) dissolve in the oxidation zone of the deposit with $^{234}U/^{238}U = 1$. The dissolution is promoted by acid sulphate solutions formed in the oxidation of the sulphates that usually accompany uranium mineralization. There is a direct correlation between the content of sulphides in an ore and the value of $^{234}U/^{238}U$. In publications on this subject, particularly those on uranium ores, the notation $^{234}U/^{238}U = \gamma$ is accepted [5.16, 5.21].

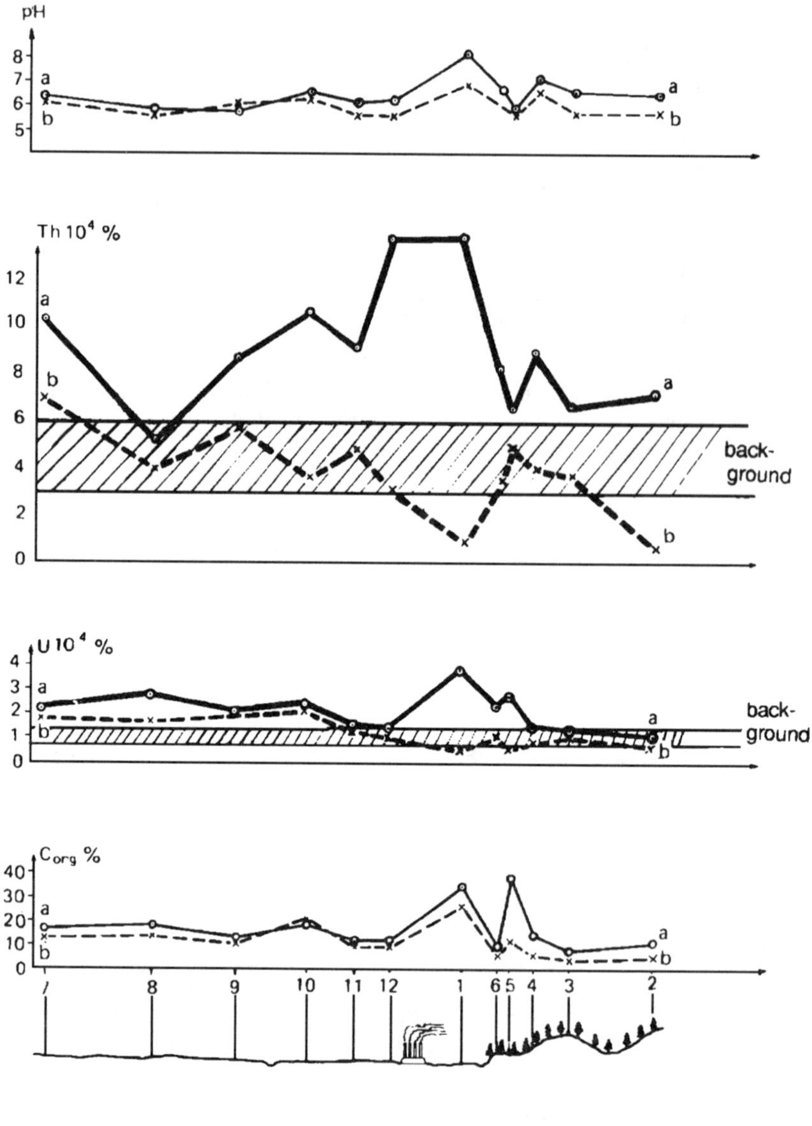

Fig. 5.12. The values of pH, Th, U and organic carbon (C_{org}) in the soil around coal-fired power station that has been operating for more than twenty years. The overall profile length being 15 km. The sampling depth being: *a*—0-3 cm, *b*—8-12 cm, 1-12—sampling points

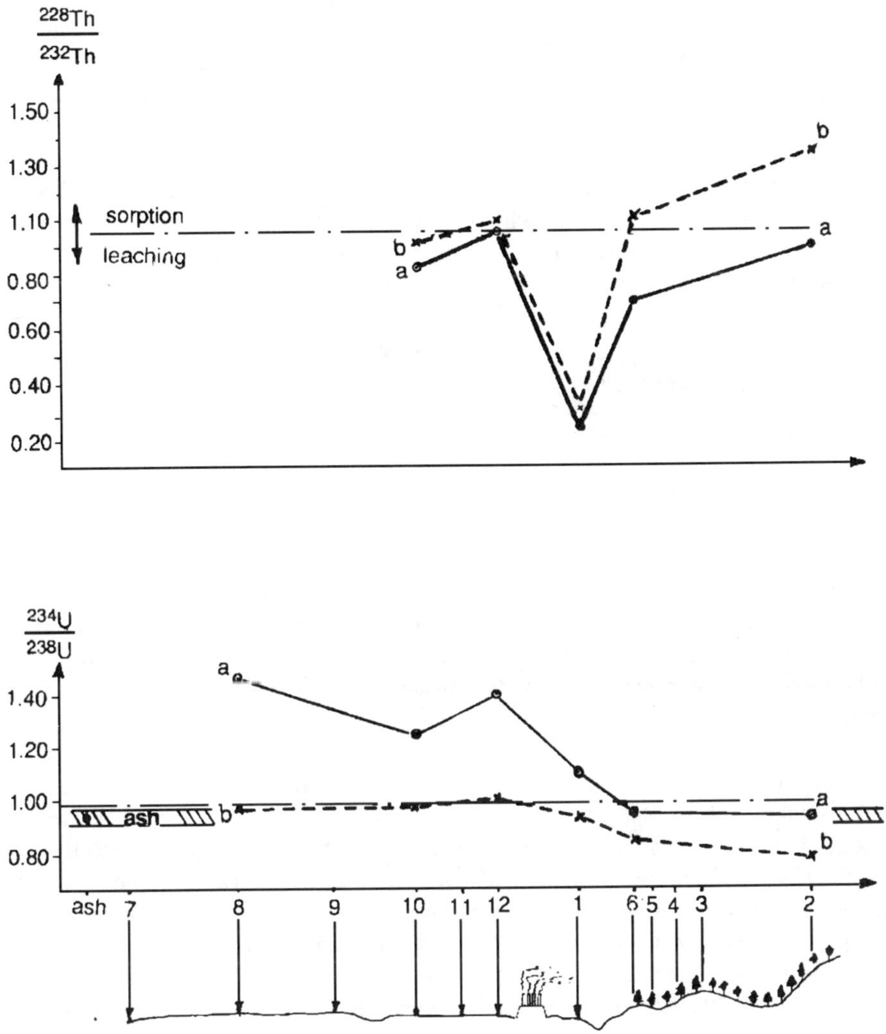

Fig. 5.13. Distribution of isotopic ratios of U and Th in soil around coal-fired power station that has been operating within more than twenty years. For notation, see Fig. 5.12

The near-equilibrium isotopic relation of uranium in underground water in combination with a higher overall uranium content than the background is a qualitative exploratory feature of uranium mineralization and is employed in hydrochemical uranium prospecting. More common is a lower than the background value of $^{234}U/^{238}U$, which is characterized by the water halo of dispersion round the orebody. The main influence on the isotopic composition of uranium γ_h and its concentration C_h in the water of the halo is the mixing of the two components: of the ore one with c_o and γ_o and the background with C_b and γ_b. Another process which can affect the composition is isotopic exchange, but, in most cases, it is of minor importance. The isotopic composition of uranium in the halo will be [5.21]:

$$\gamma_h = \frac{\gamma_o C_o (C_h - C_b) + \gamma_b C_b (C_o - C_b)}{C_h (C_o - C_b)} \tag{5.5}$$

6. *The Study of Anthropogenic Migrations of Uranium and Thorium*

The isotopic disequilibrium method can best be used to study the migration of natural radionuclides both in natural systems, and in systems that have appeared due to industrial activity (technogenic anomalies, Ch. 8). If a technogenic anomaly contains higher than background concentrations of U and Th, then the emergence of conditions for a disturbance of the radioactive equilibrium in the natural decay series is probable. The isotopic ratios $^{234}U/^{238}U$ and $^{228}Th/^{232}Th$ in this case can be indicators of the intensity of migration. The time of accumulation of measurable quantities of ^{234}U recoil atoms takes a few tens of thousands of years. At the same time ^{228}Th recoil atoms will be present in several months' time. Therefore, the $^{228}Th/^{232}Th$ ratio can almost always be employed as an indicator of migration. The $^{234}U/^{238}U$ ratio can only be applied where we have contamination by uranium-bearing minerals older than 40 thousand years (uranium isotopes have not been homogenized).

As a rule, technogenic anomalies deviate more prominently from the radioactive equilibrium than natural systems do. The reasons for this are the following. First, pollutants, besides U and Th, usually contain high concentrations of macroelements, including anionogenic ones. Waters from the precipitation permeating these substances are enriched with macrocomponents and become chemically more active. Interacting with particles of contaminants, they will primarily leach out the weakly bound recoil atoms. Second, the technogenic pollution on the surface of background soils creates a concentration gradient, which promotes an increase in the rate of the diffusion of recoil atoms.

The higher the gradient, the greater the deviation from the equilibrium will be.

Example. A study of soils around a coal-fired power station in a river valley revealed the following distribution of radionuclides. In the topmost horizon of the underlying soils (0-2 cm) the distribution of U and Th is at a maximum value in the proximity of the station. At a depth of about 10 cm the distribution of elements is the mirror image. The $^{228}Th/^{232}Th$ ratio is smallest near the station (0.25-0.32). 7-10 km away from the station the concentrations in both horizons are equal and correspond to the background, as is the $^{228}Th/^{232}Th$ (0.85-0.98) (Figs 5.12, 5.13).

These figures show that near the power station the surface soils are contaminated by U and Th as a result of the fallout of fly ash. At the same time the emission of acid gases (SO_2, CO, CO_2) during the burning of coal promotes increase of the concentration of sulphate and carbonate ions and the leaching of various metals along the whole soil section. Leaching leads to the removal of U and Th from layers with background concentrations and the formation of a minimum on the distribution curve in the horizon. Leaching also influences the fractionation of thorium due to the removal of ^{228}Th recoil atoms. The concentration gradients between the layers of soils maintain the diffusion of ^{228}Th in the underlying horizons [5.2].

5.5.2 Systematics of the $^{230}Th/^{232}Th$ Isotopic Ratio in Contemporary Volcanic Rocks

The isotopic ratios of radioactive elements can also be used as tracers of endogenic processes by analogy with the ratios of stable radiogenic isotopes. However, the application of radioactive isotopes is restricted in time by the relatively short life of the daughter radionuclides and does not go beyond the Quaternary period. Therefore, of the endogenic processes only recent volcanism can use radioactive isotopes as natural tracers.

When studying a volcanic eruption, relatively short-lived radionuclides are frequently used; e.g., $^{228}Th/^{228}Ra$, $^{222}Rn/^{226}Ra$, $^{226}Ra/^{238}U$, $^{210}Po/^{210}Pb$. The petrogenesis of volcanic rocks is studied with the aid of radionuclides which have substantially longer half-lives.

In order to study the origin of volcanic rocks and to shed more light at their regional source, the ratio of the most long-lived isotopes of thorium $^{230}Th/^{232}Th$ is used. This ratio involves thorium isotopes which belong to different decay series: ^{230}Th being a decay product of the ^{238}U series, while

^{232}Th is the parent of the thorium series. The thorium isotopic method [22-27] can be applied to rocks no older than a few hundreds of thousand years. During the formation of a volcanic rock there is no fractionation of the ^{230}Th and ^{232}Th isotopes. The change in their isotopic ratio is only possible due to two factors:

(1) the contamination of the melt by a substance with different thorium isotope composition, and

(2) the radioactive decay of unsupported ^{230}Th or its accumulation from excess uranium when the time passed from the moment the rock had formed exceeds the first few ten thousands of years, while the radioactive equilibrium between ^{238}U and ^{230}Th is disturbed. Thus, in recent or Holocene volcanic rocks the $^{230}Th/^{232}Th$ ratio will reflect the isotopic composition of thorium in the magmatic chamber.

The $^{230}Th/^{232}Th$ ratio in the chamber and the regional source of the magma will only coincide when during volcanic process, i.e. beginning from partial melting followed by evolution of the melt up to the eruption, there was no separation of ^{238}U which yields ^{230}Th and ^{232}Th.

Numerous analyses of samples taken during eruption make it clear that frequently the ratio between ^{234}U (which is always in equilibrium with ^{238}U) and ^{230}Th does not correspond to radioactive equilibrium [5.16]. Thus we can assert that the disturbance of the equilibrium between ^{230}Th and ^{234}U (the separation of U and Th) occurs at the very last stage of the volcanic process, when the melt boils and the volatile elements separate into a gaseous phase [5.17]. Now, if this is actually the case, then the isotopic composition of thorium in the chamber reflects the composition of the source, since throughout the evolution of the melt from the source to the volcanic chamber ^{234}U and ^{230}Th are in radioactive equilibrium with ^{238}U. Thus we may use the value of $^{230}Th/^{232}Th$ to estimate the thorium-uranium ratio in the chamber and in the regional source, which is usually noted by $(Th/U)_{Th}$ or K_{Th} (by analogy with K_{Pb}), and expressed in mass units as

$$K_{Th} = (Th / U)_{Th} = \frac{\lambda_{238} / \lambda_{232} \times \left(1g^{232}Th / 1g^{238}U\right)_A}{\left(^{230}Th/^{232}Th\right)_A} \times \frac{238}{232}$$

where λ_{238} and λ_{232} are the decay constants of ^{238}U and ^{232}Th, respectively, the subscript A indicates the activity units; and 238 and 232 are the mass numbers of the radionuclides.

Fractional crystallization does not separate the uranium from thorium, for which reason $^{230}Th/^{232}Th$ in comagmatic series rocks is constant [5.23].

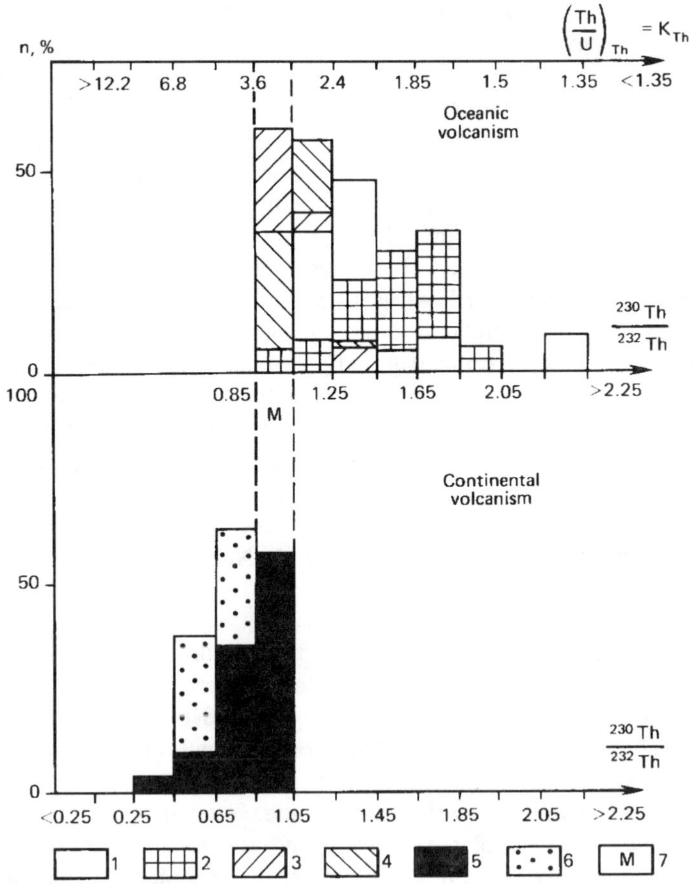

Fig. 5.14. Histograms of the distribution of $^{230}Th/^{232}Th$ relation of activity and K_{Th} in Quaternary volcanic rocks of the world:

1—MOBR; *2*—islands arcs; *3*—Iceland; *4*—oceanic islands; *5*—continental rifts; *6*—Major Caucasus (Elbrus, Cazbek); *7*—mantle region

In Fig. 5.14 we show a thorium isotopic systematics in the form of histograms. The values of $^{230}Th/^{232}Th$ and, consequently, $(Th/U)_{Th}$, fall into three groups, which belong to the three basic reservoirs on the Earth: the uncontaminated upper mantle, the continental lithosphere, and the oceanic lithosphere altered. We mentioned above that partial melting does not substantially change the thorium-uranium ratio relative to the natural fluctuations of this ratio. Therefore, the depleted and non-depleted zones of the upper mantle will not differ in the isotopic composition of the thorium and thorium-uranium ratios. The 'mantle' part of the histogram exhibits relatively constant values of $^{230}Th/^{232}Th \cong 0.95$ and $(Th/U)_{Th} \cong 3.5$. The 'oceanic' and 'continental' groups of thorium isotopic ratios are complementary to the mantle values. The 'oceanic' group includes basalts from the ocean floor, mid-oceanic ridges, and volcanic rocks of oceanic islands and young island arcs. All the rocks of this group have ratios $^{230}Th/^{232}Th > 1.0$, which are higher than the mantle ones, and, consequently, lower values of $(Th/U)_{Th} \leq 3.0$.

The maximum ratios $^{230}Th/^{232}Th$ (10-25) were obtained for hydrothermally changed rocks on the ocean floor. The 'continental' group contains derivatives of the sialic crust and the peripheral zones of continental rifts, which are to some extent contaminated by sialic material. The volcanites of this group, as distinct from the 'oceanic' group, are characterized by lower ratios $^{230}Th/^{232}Th \leq 0.85$ as compared to the mantle and, thus, by higher values $(Th/U)_{Th} \geq 4.0$. Naturally, we can always find intermediate values corresponding to contaminated mantle melts. Thus, the systematics of the isotopic composition of thorium is very much the same as that for strontium and niodymium. We can speak of a direct correlation between $^{143}Nd/^{444}Nd$ and $^{230}Th/^{232}Th$ and an indirect correlation between $^{87}Sr/^{86}Sr$ and $^{230}Th/^{232}Th$ for rift volcanic rocks and the rocks of 'hotspot' [5.25; 5.29].

A geochemical model has been suggested for the thorium-isotopic systematics [5.17, 5.22-5.27]. According to this model melts from the primarily mantle one, though uncontaminated and not varied by secondary processes have a constant ratio $(Th/U)_{Th}$, which indicates the homogeneous distribution of the thorium-uranium ratio in the mantle. The partial melting can change the Th/U ratio, however, it remains within the 'mantle' interval in the histograms of Fig. 5.14. Hence, the undepleted and depleted zones of the upper mantle will not differ by this geochemical index, which thus characterizes the upper mantle reservoir.

A marked change in the thorium-uranium ratio as a result of the separation of these elements will occur in reservoirs where the change in the oxidation number of uranium and the independent migration of U(VI) with respect to Th(IV) are possible. Within the continental crust uranium will be mobilized and redistributed with its consequent dislocation into the upper layers of the crust, due to metamorphic, hydrothermal-metasomatic and exogenic processes and from the upper layers it will be partially removed beyond the continents, whereas the behaviour of thorium is relatively indifferent. This will lead to an increase in the thorium-uranium ratio over time.

Conversely, in the oceanic crust and the upper part of the oceanic mantle, solid rocks interact with oceanic water which circulates in fissure zones. In the weakly alkaline oceanic water the redox reaction in the uranium-iron system will shift to the right (Ch. 3), viz.,

$$\underset{\text{in water}}{U^{6+}} + \underset{\substack{\text{in}\\\text{unchanged}\\\text{rock}}}{2Fe^{2+}} \rightleftarrows \underset{\substack{\text{in altered}\\\text{rock}}}{2Fe^{3+} + U^{4+}}$$

This will constantly lead to the oxidation of the iron at the surface of the fissures and the enrichment of these regions with uranium. That this process occurs is confirmed by investigations of core samples from deep-water drilling and the track method [5.28]. We can also see a regular increase in the concentration of uranium and the oxidation of iron in proportion to the age of the rock and the distance from the spreading axis.

An example of the use of thorium isotopic systematics is the analysis data of volcanic rocks in the Kuril-Kamchatka arc [5.26]. The ratio for contemporary Kamchatka volcanites $(Th/U)_{Th} = 1.6 \pm 0.3$ (for practically all the active volcanoes). In the histogram of Fig. 5.13 the isotopic compositions of Kamchatka rocks are at the extreme left. They overlap partially with tholeiites from mid-oceanic ridges, though on the whole they are displaced towards hydrothermally altered oceanic volcanites. This kind of volcanic rock composition in Kamchatka (as well as the main Kuril arc) makes it possible to refer them to derivatives of the oceanic upper mantle. However, their distinction from the basalts of mid-oceanic ridges and from oceanic islands shows a different genesis and calls for interpretation. Contemporary petrological ideas [5.30] indicate that the prerequisite for volcanic processes in island arcs, beginning with the isolation of the primary magma and ending with the formation of differentiated volcanic series, is the presence of fluid, generated from the Beniof zone. In addition during the formation of the most widespread calc alkali series on island arcs, the aqueous composition of fluids is substantially oxidized. These properties distinguish them from the fluids from mid-oceanic ridges, which are 'drier' and less oxidized. If the role of aqueous oxidized fluid is really decisive in the formation of island-arc volcanites, then it should influence the behaviour of uranium, and consequently affect the $(Th/U)_{Th}$ and $^{230}Th/^{232}Th$ ratios. This kind of fluid can oxidize the uranium in enclosing rocks, extract and displace it into the magma formation zone, thus enriching the melt with uranium. If this occurs, then the volcanic rocks of the Kuril-Kamchatka arc will have a direct correlation between $^{230}Th/^{232}Th$ and U distinct from the tholeiites of the mid-oceanic ridges. This is exactly what we see (Fig. 5.14). Hence, the isotopic geochemical pattern does confirm the petrological data.

An interesting study is the comparison of $(Th/U)_{Th} = K_{Th}$ with K_{Pb} and the common elemental thorium-uranium ratio K. K_{Th} and K_{Pb} reflect the distribution of the Th/U ratio at different stages of the evolution of the substance. K_{Pb} represents an integral value of this ratio in the preoceanic period (longer than 2×10^8 years) of the development of the source. K_{Th} is the reflection of the contemporary distribution of the thorium-uranium ratio in the source. The value K refers directly to the rock and is subjected to the influence of all secondary processes, beginning with the displacement of melt along the volcanic channel. Therefore, K_{Th} is the more reliable parameter than K and characterizes the primary melt, K_{Pb} is not equal to K_{Th} since it is to a great extent a convenient value. However, a comparison of the systematics of K_{Th} and K_{Pb} makes it possible to establish the following patterns common for K_{Th} and K_{Pb}:

(1) The homogeneity of K_{Pb} for various types of oceanic rocks independent of their genesis makes it possible to classify them as derivatives of a single homogeneous mantle source substantially distinct from the sialic crust. Rocks of this kind also include volcanites of the Kuril-Kamchatka arc.

(2) Markedly 'continental' values of K_{Th} and K_{Pb} for individual zones of the ocean in the Southern Hemisphere make it possible to speak of traces of ancient continents in these regions.

(3) The emergence of K_{Th} and K lower than mantle values in contemporary volcanic rocks of the ocean and the absence of these distinctions for K_{Pb} prove that K_{Th} and K are affected by relatively young processes, no older than the time of formation of the rocks of the present-day ocean.

REFERENCES

5.1. Titayeva, N.A., Veksler, T.J., Orlova, A.V., 1977. Radium in Recent Volcanic Rocks in Kamchatka.—Geologiya i razvedka. Izv. vysh. uchebn. zav., No 4, 70-76.

5.2. Titayeva, N.A., 1992. Geochemistry of the Nature Radionuclides in Hypergenic Zone.—The Problems of Radiogeochemistry and Kosmochemistry. Nauka, Moscow, 64-129.

5.3. Titayeva, N.A., 1980. Radioactive Isotopes in Calcium Chloride Brines. Dokl. Acad. Nauk SSSR, **251**, No 4, 974-976.

5.4. Titayeva, N.A., Taskaev, A.I., 1983. Migration of Heavy Natural Radionuclides in the Conditions of Humid Zone. Nauka, Leningrad.

5.5. Serdukova, A.S., Kapitonov, Yu.T., 1975. Isotopes of Radon and their Decay Products in Nature. Atomizdat, Moscow.

5.6. Ermolayeva-Makovskaya, A.P., Litver, B.Ya., 1978. ^{210}Pb and ^{210}Po in Biosphere. Atomizdat, Moscow.

5.7. Osmond, J.K. and Cowart, J.B., 1982. Natural Uranium and Thorium Series Disequilibrium: new approaches to geochemical problem.—Nuclear Science Application, 305-352.

5.8. Baturin, G.N., 1975. Uranium in the Recent Marine Sedimentation. Atomizdat, Moscow.

5.9. Arslanov, Kh.A., Tertichny, N.I., Gerasimova, S.A., Lokshina, N.B., 1976. To the Question of Sea Mollusk Shells Dating Relative to $^{230}Th/^{234}U$.—Geochemistry, No 11, 1724-1734.

5.10. O'Brien, Veeh, H.H., Cullen, D.J. and Milnes, A.R., 1986. Uranium-series Isotopic Studies of Marine Phosphorites and Associated Sediments from the East Australian Continental Margins.—Earth and Planet. Sci. Lett., No 1/2, 19-35.

5.11. Atkinson, T.C., Harmon, R.S., Smart, P.L., Waltham, A.S., 1978. Paleoclimatic and Geomorphic Implication of ^{230}Th/^{234}U Dates on Speleotherms from Britain.—Nature, **272**, No 5648, 24-28.
5.12. Cherdyntsev, V.V., 1969. Uranium-234. Atomizdat, Moscow.
5.13. Scholten, J.C., Botz, R., Mangini, A. et al., 1990. High Resolution ^{230}Th Stratigraphy of Sediments from High-latitude Areas (Norwegian Sea, Fzam Strait).—Earth and Planet. Sci. Lett., **101**, No 1, 54-62.
5.14. Titayeva, N.A., 1980. Some Data Concerning the Age of Quaternary Deposits by Means of Ionium Method (According to Bone Remains).—Moscow University Geology Bulletin, No 3, 51-61.
5.15. Kuznetsov, Yu.V., 1976. The Ocean Radiochronology. Atomizdat, Moscow.
5.16. Cherdyntsev, V.V., 1978. Nuclear Volcanology. Nauka, Moscow.
5.17. Titayeva, N.A., 1986. The Application of the Thorium and Uranium Isotopes Rations for the Iinvestigation of the Volcanic Processes.—The Evolution of the Crust-mantle System (ed. Yu.A. Shukolukov). Nauka, Moscow.
5.18. Titayeva, N.A., Ananiev, V.V., Orlova, A.V., Veksler, T.I., 1970. The Ionium Method in Determining the Volcanic Rock Age. Moscow University Geology Bulletin, No 4, 65-69.
5.19. Cochran, J.K., McKibbin-Vonghan, T., Dornblaster, M.D. et al., 1990. ^{210}Pb Scavenging in the North Atlantic and North Pacific Oceans.—Earth and Planet. Sci. Lett., **97**, No 3/4, 332-352.
5.20. Zverev, V.L., Tokarev, A.N., Tyminskii, V.G., and Shvetz, V.M., 1980. Radioisotopic Geochemistry. Nedra, Moscow.
5.21. Syromyatnikov, N.G., 1961. The Migration of Uranium, Radium and Thorium Isotopes and the Interpretation of Radioactive Anomalies. Acad. Sci., Publ. of Kazakh SSR, Alma-Ata.
5.22. Titayeva, N.A., 1990. The Geochemical Tracer Role of Isotopic Data of Thorium-uranium Relation in Volcanic Rocks.—Acad. Sc. Report, **312**, No 6, 1554-1557.
5.23. Titayeva, N.A., Polyakov, A.I., Orlova, A.V., Zornina, Yu.B., 1982. Thorium-uranium Ratio as Tracer of Comagmatism of Volcanic Rhyolite-Basalt Series in Iceland.—Geochemistry, No 11, 1632-1638.
5.24. Titayeva, N.A., 1983. The Isotopes of Uranium and Thorium in Products of the Recent Volcanism.—The Problems of Radiogeology (ed. N.P. Laverov and G.B. Naumov). Nauka, Moscow, 163-172.
5.25. Polyakov, A.V., Titayeva, N.A., Orlova, A.V., Zornina, Yu.B., 1986. The Isotopic Composition of Thorium in Volcanic Rocks of Continental Rift Zones.—Geochemistry, No 5, 612-619.
5.26. Titayeva, N.A., Orlova, A.V., Karpushina, T.I., Nikulin, V.I., 1989. The Isotopic Composition of Thorium in Quaternary Volcanic Rocks of Kuril-Kamchatka Arc and Some Questions in Petrogenesis.—Geochemistry, **7**, 973-983.
5.27. Titayeva, N.A., 1982. A New Isotope-geochemical Tracer in Petrogenesis of Volcanic Rocks.—USSR Acad. Sc. Reports, **263**, No 5, 1221-1224.
5.28. Aumento, F., 1979. Distribution and Evolution of Uranium in the Oceanic Lithosphere.—Philos. Trans. Roy Soc., London, I.S.S.N. 0080-4614, **291**, No 1381, 423-432.
5.29. Condomines, M., Hemond, Ch. and Allegre, C.J., 1988. U-Th-Ra Radioactive Disequilibria and Magmatic Processes.—Earth and Planet. Sci. Lett., **90**, 243-262.
5.30. The Origin of Volcanic Island Arc Series, 1985. (Ed. T.I. Frolova). Nauka, Moscow.

6 Radioactivity and the Earth's Heat

The source of the Earth's thermal energy has long been debatable and remains so. The increase in temperature with the depth as recorded in mines and boreholes and volcanic activity indicate the existence of high temperatures in the interior of the Earth.

The Earth's thermal energy is the driving force for many, mainly endogenous, geological processes, such as the formation of the nucleus, the evolution of crustal mantle system, the degassing of the Earth with the appearance of the ocean and the atmosphere, the movement of the lythospheric plates, mountain building, volcanism, and the formation of mineral deposits. A knowledge of the distribution of the Earth's heat over time and space will indicate how the geodynamic evolution of the Earth proceeded. The thermal history of the Earth can be reconstructed from (1) the initial distribution of temperatures; (2) the distribution of the sources of heat; (3) the thermal field on the Earth's surface at present; (4) the change in temperature with the depth; (5) the change in thermal and physical properties with the depth (thermal capacity, thermal conductivity, etc.); (6) the mechanism of heat transfer. Most of these parameters are either unknown or are known only tentatively or partially (for instance, the distribution with depth).

6.1 The Contemporary Thermal Field of the Earth

The basic empirical parameter characterizing the Earth's contemporary thermal field is the heat flow. The density of the Earth's heat flow (q) expresses the losses in thermal energy by the Earth's unit surface per unit time, and is usually called the heat flow. The heat flow is the product of geothermal gradient along the normal to the Earth's surface (grad T_n) and the thermal conductivity of the medium (k):

$$q = -k \text{ grad } T_n \tag{6.1}$$

The unit of heat flow (HFU) is 10^{-6} cal/cm^2·s. The geothermal gradient on land is determined by measuring temperatures when drilling deep holes, and

Table 6.1 The Distribution of Heat Flow on the Earth's Surface [6.2]

	Number of measurements	Average heat flow	
		HFU	mW/m^2
	Continents		
Archean	89	1.00	41.87
Proterozoic	153	1.12	46.89
Phanerozoic-neorogenic regions	206	1.29	54.01
Early Paleozoic orogenic regions	54	1.37	57.36
Late Paleozoic orogenic regions	452	1.46	61.13
Mesozoic orogenic regions	67	1.57	65.74
Tertiary orogenic regions	478	1.77	74.11
	Oceans		
Ocean basin	1822	1.47	61.55
Spreading centres	1145	2.63	110.12
Ocean trenches	63	1.16	48.57
Continental marginal areas	193	1.78	74.53
Continental slope	141	1.81	75.78
Aseismic regions	205	1.74	72.85
Fractured zones	59	2.06	86.25
Island arcs	89	1.71	71.60

1 HFU = 1×10^{-6} cal/cm^2·s = 41.87 mW/m^2
1 HGU = 1×10^{-13} cal/cm^3·s = 0.4187 mW/m^2

in the ocean, during deep-water drilling

$$\text{grad } T_n = dT/dZ \tag{6.2}$$

where T is the temperature, and Z is the depth. Measurements taken on land should be below the neutral layer, which separates the layer where there are seasonal fluctuations of temperatures from the deeper layers. Grad T_n on land varies from 0.5 °C to 7 °C per 100 m. Where there are local thermal anomalies it can change from negative values to more than 100 °C within 100 m (e.g. the exit of thermal sources, volcanoes, technogenic anomalies, etc.). In the case of sedimentary layer in the ocean it varies from 1.5 °C to 15 °C per 100 m [6.1].

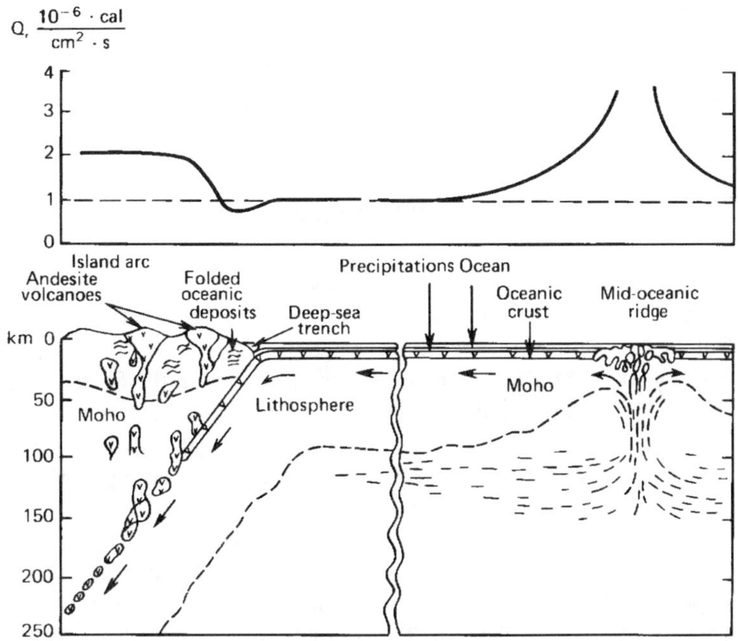

Fig. 6.1. A generalized model of heat flow associated with the spreading of ocean floor. Maximum heat flow in mid-oceanic ridge and its minimum variety in the deep-water trench where subduction takes place (according to Blackwell, 1971)

The thermal conductivity of rock is estimated in laboratory conditions. It ranges from 1.5-3.0 × 10^{-3} cal/cm·s·°C for bottom sediments to 7.0-10.0 × 10^{-3} cal/cm·s·°C for acid and ultrabasic igneous rocks; 10-15 × 10^{-3} cal/cm·s·°C for halogen rocks and quartzites.

The heat flow q varies less than k and grad T_n; it varies between 0.7 HFU and 3.0 HFU (Table 6.1). The distribution of heat flow is associated with the geological and tectonic conditions of the terrain. According to Smirnov [6.1], each homogeneous genetic series of tectonic structures is characterized by the same heat flow, which in quantitative terms reflects their contemporary energetic condition. The heat flow distributions in stable and active tectonic zones are markedly different. Stable regions have a homogeneous distribution of heat flow. Small q are characteristic of Precambrian platforms, Phanerozoic platforms have a larger heat flow, and it is still higher where there is Phanerozoic folding, being increased from ancient structures to those that are younger.

In tectonically active zones the heat flow changes markedly. The largest q are found in rift structures and hotspots; the smallest are in deep-water trenches (Fig. 6.1). The heat flow for the Alpine and for contemporary geosynclines differ with the latter being somewhat lower.

6.2 Sources of the Earth's Heat

There are two main sources of thermal energy on Earth: (1) the heat generated due to radioactive decay (radiogenic heat); (2) the initial heat, which has remained in the Earth since its formation.

Radiogenic heat flow occurs due to the presence in the Earth of radionuclides (Table 6.2). The evolution of thermal energy takes place as a result of nuclear transformations and interactions between charged particles, recoil nuclei, gamma quanta, and fission fragments and the surrounding atoms. The amount of heat released due to radioactive decay 1 MeV = 1.6×10^{-6} erg = 3.83×10^{-14} cal. The sources of radiogenic thermal energy of the Earth are radionuclides of natural decay series and long-lived radionuclides that do not give any series. Over 99% of the radiogenic heat flow is due to the decay series and ^{40}K. Even though the distribution of ^{87}Rb is higher than ^{40}K, it emits low-energy β-particles (44 KeV), which only create small portion of the radiogenic heat. The contribution of other radionuclides is even smaller.

If we know the composition of rocks in an envelope of the Earth and the content of U, Th and K, then the average heat (A) generated in the envelope can be estimated according to the formula

$$A = 0.317\rho\ (0.73\text{U} + 0.20\text{Th} + 0.27\text{K})$$

Table 6.2 Generation of Radiogenic Heat from U, Th, and ^{40}K [6.3]

Series, Element	MeV/atom	W/kg	J/g·year	cal/g	cal/year
Series ^{238}U	47.4	9.62×10^{-5}	2.97	2.30×10^{-8}	0.71
Series ^{235}U	45.2	58.6×10^{-5}	17.0	14.0×10^{-8}	4.3
				0.63×10^{-8}	0.20
Series ^{232}Th	39.8	2.63×10^{-5}	0.84	0.67×10^{-8}	0.20
^{40}K	0.71	2.80×10^{-5}	0.92		

where U and Th are the respective element concentrations in g/t; and K is the content in %, ρ is the density of the rocks in g/cm³. The unit of heat generation is HGU, i.e. 1 HGU = 1×10^{-13} cal/cm³/s.

The generation of radiogenic heat by individual layers and shells, and by the Earth as a whole depends on the chosen model. In Table 6.3, we present the calculation by Smyslov [6.3]. Smirnov's data [6.1] suggests that the distribution of radioactive heat sources is somewhat different, viz., the sedimentary layer yields 3.4 ± 0.6 HGU; the crystalline rocks in ancient platforms yield 2.8 HGU; young platforms yield 3.5 HGU; the foredeeps yield 2.7 HGU; the granitic layer yields 3.7 ± 0.7 HGU; the basaltic layer yields 1.1 ± 0.5 HGU; and the upper mantle yields 0.02 ± 0.01 HGU.

The mantle heat flow (q_m) is defined from the difference between the estimated heat flow (q) and the radiogenic heat flow (q_r). The mantle heat flow is the part of the overall heat flow supplied from the Earth' depth through the Mohorovicic discontinuity. Mapping of the mantle heat flow has shown that it varies from less than 0.4 HFU in Precambrian structures to 2.5-3.0 HFU

Table 6.3 Generation of Radiogenic Heat by Various Layers of the Earth's Crust [6.3]

Layers of the Earth's crust	Thickness, km	Generation of radiogenic heat 10^{-6} W/m²	Density of heat flow 10^{-2} W/m²	
			radiogenic component	observation average
		Continents		
Sedimentary	6	1.2	0.36	
Granitic and metamorphic	12	2.2	1.32	—
Dioritic	12	1.1	0.66	—
Basaltic	18	0.35	0.31	—
The Earth's continental crust (taken as a whole)	48	—	2.65	5.52
		Oceans		
Sedimentary	2	2.0	0.2	—
Dioritic	6	1.1	0.3	—
Basaltic	6	0.2	0.01	—
The Earth's oceanic crust (taken as a whole)	14	—	0.56	5.92

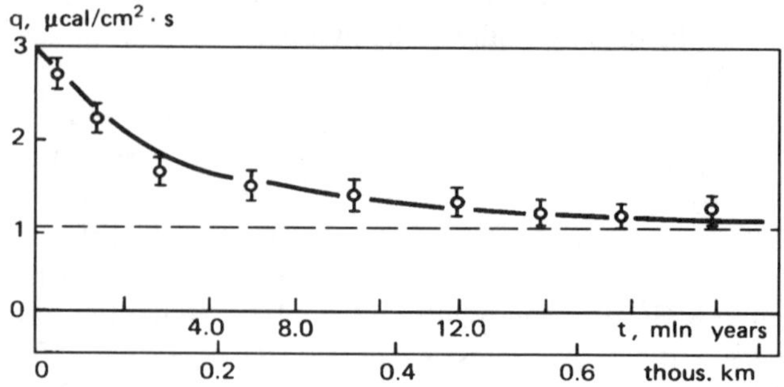

Fig. 6.2. Dependence of heat flow on the age of ocean floor (according to E.A. Lyubimova, 1983)

in the depressions of the marginal seas [6.1]. Tectonically active belts correspond to a high thermal potential of the upper mantle. It is here that the thermal energy that comes from the mantle is 5-7 times greater than the average heat flow of the Earth.

The relation between the radiogenic and mantle components of the heat flow in structurally different tectonic zones is not the same. For most of the blocks of the Earth's continental crust with a complete set of layers, the contribution of radiogenic heat makes up 40% or more. For ancient stabilized platforms (the Russian, Siberian, and North-American ones) the figure is as high as 60%. Conversely, as distinct from areas of stationary thermal conditions, the basic contribution from tectonically active regions that have nonstationary thermal field is the mantle component, which reaches here 80%. Within the earth's oceanic blocks and in the transition zone from the continent to the ocean the radiogenic component does not exceed 10-20% [6.3]. In spreading zones, where hot mantle material is uplifted, the mantle flow is the highest. It diminishes smoothly with distance from the axis of mid-oceanic ridge, i.e. as the age of the ocean crust increases [6.3]

$$\bar{q}_m \approx \frac{1}{\sqrt{\tau}}, \qquad (6.3)$$

where τ is the age in millions of years (Fig. 6.2).

In those layers of the Earth's crust where the radiogenic thermal flow is predominant, the temperature at any depth can be estimated by the equation:

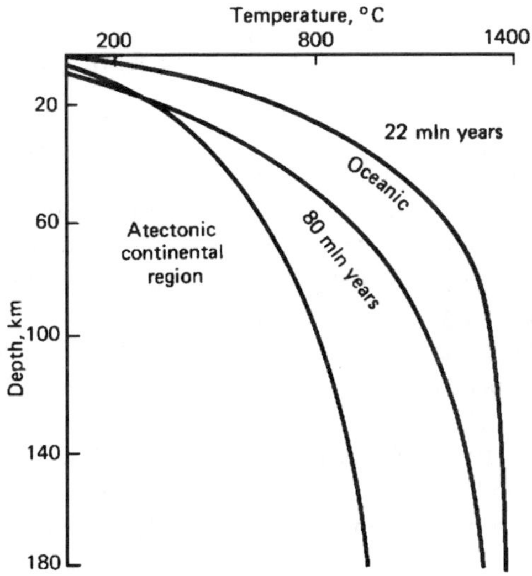

Fig. 6.3. Continental and ocean geotherms (according to Oxburgh, 1980)

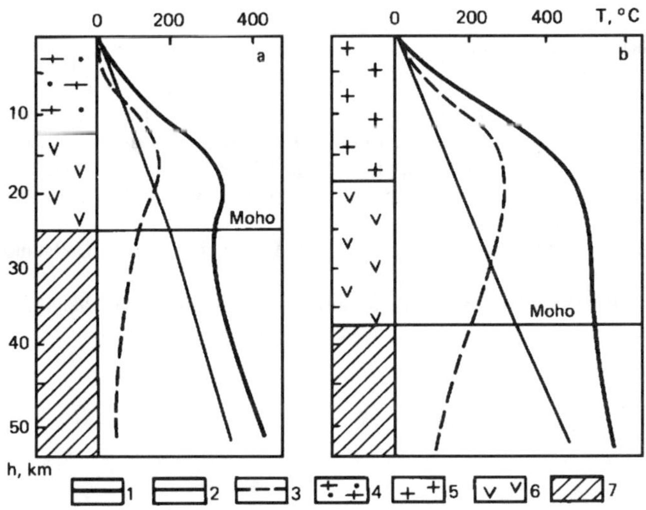

Fig. 6.4. The zone of the change of geothermal gradients in the earth's crust, conditioned by the contribution of radiogenic constituent of granitic and metamorphic layer. The blocks of the earth's crust:

a–deeply eroded (granulitic belt of the Kola peninsula); *b*–with complete profile of radiogeological zones and a thick granitic and metamorphic layer (the Baikal region). *1*–the summation geotherms; *2*–the mantle component; *3*–radiogenic component; the layers of the lithosphere: *4,5*–weakly radioactive granitic and metamorphic; *6*–granulitic and basic; *7*–subcrustal, upper mantle (according to A.A. Smyslov, 1983)

$T = T_0 + (q/k)Z - AZ^2/2k$, where T is the temperature (in °C), T_0 is the temperature on the Earth's surface (in °C); q is the surface heat flow (μ·cal/cm·s); Z is the depth (km); A is the rate of heat generation (10^{-13} cal/cm·s); and k is the thermal conductivity in μ·cal/cm·s °C. By calculating the temperatures at particular depths for various points on the Earth's surface, it is possible to construct geotherms, or the curves describing the distribution of temperature with depth. Figure 6.3 gives an example of a continental geotherm for q = 1.2 HFU and two oceanic geotherms that correspond to the oceanic rocks of different age [6.4]. The investigation of the Earth's heat flow has led to a class of models of the Earth's thermal history where the radiogenic component of the heat flow largely depends on the original ratios K/U and Th/U (Fig. 6.4).

The Earth's thermal energy in places of high thermal flow (≥ 2HFU) is often transformed into the energy of volcanism and hot springs. In many such places the geothermal energy can be employed practically as heat sources and to generate electricity (Iceland, Japan, Kamchatka (Russia), New Zealand, Italy, and the USA).

REFERENCES

6.1. Smirnov, Ya.B., 1980. The Thermal Field of the USSR Territory (an explanatory note of the map: The Heat Flow and Subsurface Temperatures, scale 1 : 10 000 000). The Main Board of Geodesy and Mapping at the USSR Council of Ministers, Moscow.

6.2. Aswathanarayana, U., 1985. Principles of Nuclear Geology. Oxonian Press Pvt. Ltd., New Delhi.

6.3. The Problems of Radiogeology, 1983. (Ed. N.P. Laverov and G.B. Naumov). Nauka, Moscow.

6.4. Oxburgh, E.R., 1980. Heat Flow and Magma Genesis.—Physics of Magmatic Processes (Ed. Hagreaves), Princeton Univ. Press, 161-199.

7 Radionuclides as the Products of Natural Nuclear Reactions

'The penetrating radiations discovered by Gloecker and Hess come from upper sources and possess the energy of the power, the origin of which on Earth is hardly probable, ... they incessantly permeate and cover the biosphere (presumably performing in its matter changes of yet unknown magnitude.'

V.I. Vernadsky

7.1 Nuclear Reactions

In a narrow sense, a nuclear reaction takes place when a simple or complex microparticle collides with a nucleus. In these conditions the nucleus is referred to as the target. A nuclear reaction can be written as a chemical reaction

$$a_1 + a_2 \rightarrow b_1 + b_2 + \ldots b_j$$

where a_1 and a_2 are the reacting particles and b_1, b_2, ... b_j are the particles which emerge as a result.

The first nuclear reaction was carried out by Rutherford in 1919 by bombarding nitrogen atoms by alpha particles:

$$^{14}_{7}N + ^{4}_{2}He \rightarrow ^{17}_{8}O + ^{1}_{1}H$$

The following notation is often used, $^{14}N(a, p)^{17}O$, where the bombarding particles and those that are formed stand within the brackets, while the target nucleus is written to the left and the product nucleus to the right outside the brackets.

In nuclear physics the laws of conservation of electric charge, energy and momentum are important.

The quantitative characteristic of an elementary collision is the effective cross-section, which does not depend on the flux of bombarding particles and the number of target particles. The physical meaning of an effective cross-section is the cross-section area of the effective volume round the target nucleons which the bombarding particle must enter to cause the nuclear

Table 7.1 Effective Cross-sections for the Absorption (*n*, γ) and Scattering of (*n*, *n*) of Thermal Neutrons (in barns)

Element	Absorption	Scattering	Element	Absorption	Scattering
H	0.33	81.4	Ca	0.43	3.2
Li	70.0	1.2	Mn	12.6	2.0
Be	10^{-2}	V	4.98	—	—
B	753.0	4.4	Fe	2.43	11.8
C	4.5×10^{-3}	—	Co	37.0	—
N	1.78	11.4	Ag	63	—
O	2×10^{-4}	4.24	Cd	2 450	—
F	10^{-2}	—	In	190	—
Na	0.49	3.6	Sm	6 500	—
Mg	0.06	3.7	Eu	4 500	—
Al	0.21	1.5	Gd	44 000	—
Si	0.13	2.4	Du	1 100	—
P	0.19	3.4	Hg	380	—
S	0.49	1.2	Au	98.8	—
Cl	31.6	15.0	Pb	0.17	11.4
K	1.97	2.0	Cu	3.77	—

reaction. The unit of the effective section (σ) is the barn: 1 barn = 10^{-24} cm^2 per nucleus. Effective cross-sections depend on the type of bombarding particle, its energy, and the properties of the target. In Table 7.1, we present the effective cross-sections for the reaction (*n*, γ) upon thermal neutrons (E_n = 0.025 eV).

Nuclear reactions are classified according to the type of bombarding particle as neutron, charged particle and gamma photon (photonuclear reactions) reactions.

Depending on the energy of the particles hitting the nucleus, nuclear reactions are classified as low, medium, and high energy reactions. Low energy reactions (up to several electron volts) mainly involve neutrons. Charged particle reactions are not possible in this case because by contrast to the neutrons, these particles must overcome the strong electric fields of the atom (the Coulomb barrier). Average (up to several MeV) and high energy reactions can be caused by neutrons, charged particles, and photons.

The reaction products depend on the energy acquired by the nucleus and the nature of target nucleus. If the energy received by the nucleus during the bombardment is insufficient for the emission of a nucleon, then some of the energy (or all of it) will be given away by the nucleus in the form of γ-radiation. Reactions of this kind are called radiation capture and often proceed under the action of neutrons with small energies (*n*, γ). In average energy reactions, neutrons, protons, α-particles, deutrons, and sometimes other particles can be emitted. The emission of neutrons when the energies of the reaction are relatively small is the most probable process, since neutrons do not have to overcome a Coulomb barrier. This is especially true for heavy nuclei, where the barrier is the highest. With lighter nuclei, whose Coulomb barrier is smaller, the emission of α-particles and protons is also possible; other particles are emitted by nuclei when the energies are much higher. Thus, depending on the energy of the bombarding particles, the reaction may take one of several competing courses. This can be illustrated by the bombardment of fluorine by neutrons:

$$
{}^{19}_{9}\mathrm{F} + {}^{1}_{0}n \rightarrow \begin{cases} {}^{20}_{9}\mathrm{F} + \gamma; & \text{with the neutron energy} = 1.5\ \text{MeV} \\ {}^{19}_{8}\mathrm{O} + {}^{1}_{1}\mathrm{p}; & \text{ditto} \quad 4.10\ \text{MeV} \\ {}^{18}_{9}\mathrm{F} + 2{}^{1}_{0}n; & \text{ditto} \quad 10\text{-}20\ \text{MeV} \\ {}^{15}_{8}\mathrm{O} + {}^{1}_{1}\mathrm{p} + 4{}^{1}_{0}n; & \text{ditto} \quad 20\text{-}40\ \text{MeV} \end{cases}
$$

When heavy nuclei are bombarded by neutrons, nuclear fission can occur. As a result, two nuclear fragments of unequal size and several neutrons are formed. As a rule, the fission products are radioactive isotopes of elements in the middle of the Periodic Table.

7.2 Nuclear Reactions in the Atmosphere

For nuclear reactions to be initiated, a fairly powerful flux of bombarding particles is required. Cosmic radiation, which comes onto the surface of the atmosphere from the extra-terrestrial sources, is such a flux. Cosmic radiation is classified as primary and secondary.

7.2.1 Primary Cosmic Radiation

Primary cosmic radiation consists of about 90% high energy protons, 9% helium nuclei (alpha particles), and 1% of heavier nuclei, neutrino, photons, etc. The energy spectrum of primary cosmic radiation ranges from 1 to 10^{14} MeV, and particles with the energies of 10^{19}-10^{20} MeV have been registered. The maximal intensity of radiation corresponds to energies of nearly 300 MeV. Particles with energies $\leq$ 400 MeV are generated in the Sun during solar outbursts and are called solar primary cosmic radiation. A higher energy component is galactic primary cosmic radiation. It is thought to be generated by outbursts from supernovas. Particles with energies more than 10^{12} MeV most likely emerge outside the boundaries of our Galaxy. The age of the primary cosmic radiation, i.e. the time that has passed from the generation of the particles to their arrival in the Earth's atmosphere, was calculated from the radionuclides ^{10}Be and ^{36}Cl and is estimated to be from 2.5×10^6 years to 33×10^6 years. The intensity of galactic radiation is $2 \div 4$ particles/cm^2·s. The investigation of radionuclides in meteorites has revealed that the density of galactic radiation varied not more than 2 times for the past billion years. However, within this interval there were cyclic variations.

Solar cosmic radiation with an energy of 400 MeV (1-40 MeV during solar outbursts has a very great intensity (10^6-10^8 particles/cm^2·s). Sometimes during these flashes particles of several GeV are emitted. The relation between the galactic and solar components of cosmic radiation depends on the 11-year solar cycle. The flux density of galactic radiation is minimal during the flashes, when the solar component is maximal.

The flux density of primary cosmic radiation is affected by the geomagnetic field, for which reason it is higher at the poles than on the equator.

7.2.2 Secondary Cosmic Radiation

The particles and photons produced by primary cosmic radiation and the atoms of atmospheric gases are called secondary cosmic radiation. High energy particles of primary cosmic radiation cause nuclear fission reactions with the formation of neutrons, protons, π-mesons and cations. A large number of secondary particles possess sufficient energy to induce new nuclear reactions and create new secondary particles. Such reactions are of cascade character and called showers. π-mesons decay to form μ-mesons and photons, which give rise to new showers. The original nuclear reactions with high energy particles are called activization reactions. In this connection we can speak of the formation of a large number of radionuclides. The energy of the secondary

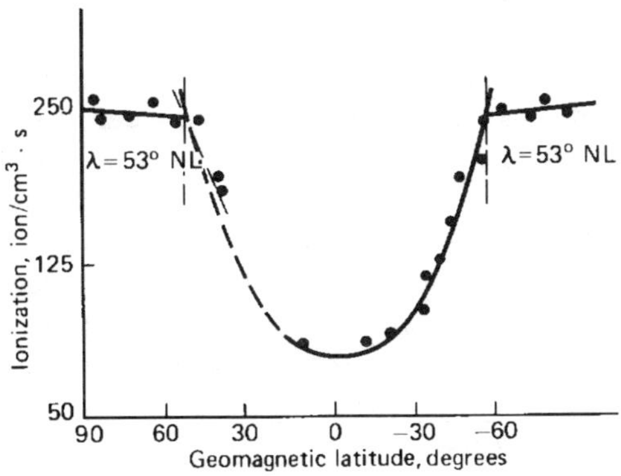

Fig. 7.1. Latitudinal effect of cosmic rays at the depth of the atmosphere, 50 g/cm^3 (in: *Natural Isotopes of the Hydrosphere*, 1975)

cosmic particles is mostly expended on ionization. The average rate of the formation of ions in the air is 2.1 ions/s·cm^3. The ionization maximum is reached during the minimum of the 11-year solar cycle. The distribution density of ions is affected by the geomagnetic field and depends on the latitude, thus increasing from the equator to the poles (Fig. 7.1). The neutron flux density has an analogous distribution. The distribution of secondary neutrons and protons in the atmosphere leads to variations in the rate of formation of cosmogenic radionuclides, which is a function of geomagnetic latitude and altitude.

The integral flux of neutrons increases with the height, and then decreases due to the decrease in the density of the atmosphere, and because the neutrons escape from the upper layers. The formation rate of neutrons averaged over the whole solar cycle is 4.0 n/cm^3·s^{-1}.

7.2.3 Formation of Cosmogenic Radionuclides in the Atmosphere

Radionuclides formed as a result of nuclear reactions induced by cosmic rays are called cosmogenic The bombarding particles in these reactions are the particles of primary and secondary cosmic radiation, while the targets are the nuclei of the atoms of atmospheric air, primarily nitrogen, oxygen, and argon.

In addition to activation reactions the most common are nuclear reactions involving protons and neutrons: (*p*, *xp*, *yn*); (*p*, *n*) (*n*, *p*); (*n*, 2*n*); (*n*, α) (*n*, γ). They yield isotopes of numerous chemical elements. The main properties of the most common cosmogenic radionuclides are presented in Table 7.2. Approximately 70% of cosmogenic radionuclides form in the upper layers of the atmosphere, and only about 30% in the troposphere. The subsequent behaviour of cosmogenic radionuclides depends on exchange processes between the stratosphere and the troposphere, the movement of air in the troposphere, and the exchange between the atmosphere and other geospheres. Cosmogenic radionuclides, with the exception of the noble gases, rapidly oxidize after their formation. Radiocarbon and tritium are preserved in the form of free oxides $^{1}H^{3}HO$ and $^{14}CO_2$. Other oxides are sorbed on aerosols and are removed from the atmosphere during condensation of moisture and subsequent removal from the atmosphere by precipitates or due to gravitational 'dry' sedimentation. Tritium comes from the atmosphere with precipitation.

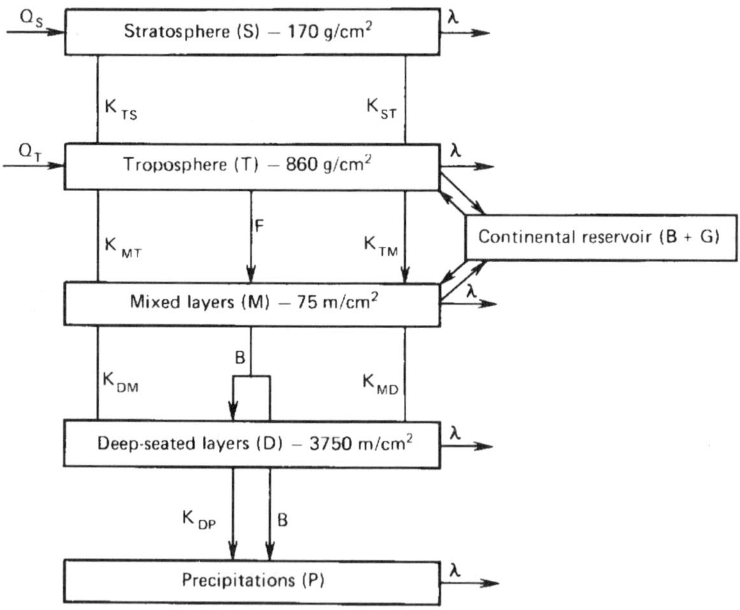

Fig. 7.2. A generalized model of exchange reservoirs of the Earth for the circulation of cosmogenic radionuclides:

Q designates the formation in the atmosphere, λ is the decay constant, *C* is the exchange constant. Other indications are given in Table 7.3 (according to Lal, 1972)

Table 7.2 Properties of the Earth's Cosmogenic Radionuclides [7.2]

Radionuclides	Main formation reactions	$T_{1/2}$	Type of decay	E_{rad}, MeV	Decay product	Formation rate, atom/cm²·s	Equilibrium amount on Earth, g
^{3}H	$^{14}N(n, t)^{12}C$	12.26 years	β^-	0.018	^{3}He	0.25	3500
	$^{16}O(p, t)^{14}C$						
^{7}Be	$^{14}N(n, 3p5n)^{7}Be$	53 days	β^-	0.48	^{7}Li	8.1×10^{-2}	3.2
	$^{14}N(p, 4p4n)^{7}Be$						
	$^{16}O(p, 5p5n)^{7}Be$						
^{10}Be	$^{14}N(p, 4pn)^{10}Be$	1.5×10^{6} years	β^-	0.55	^{10}B	4.5×10^{-2}	4.8×10^{8}
	$^{16}O(p, 5p2n)^{10}Be$						
^{14}C	$^{14}N(n, p)^{14}C$	5730 years	β^-	0.156	^{14}N	2.5	7.5×10^{7}
	$^{16}O(p, 3p)^{14}C$						
^{22}Na	^{40}Ar (split.)^{22}Na	2.6 years	β^+	0.54	^{22}Ne	8.6×10^{-5}	1.9
			γ	1.28			
^{26}Al	$^{26}Mg(p, n)^{26}Al$						
	$^{28}Si(p, 2pn)^{26}Al$	7.4×10^{5} years	β^+	1.17	^{26}Mg	1.4×10^{-4}	1.1×10^{6}
^{32}Si	^{40}Ar (split.)^{32}Si	500 years	β^-	0.1	^{32}P	1.6×10^{-4}	1400
^{35}S	^{40}Ar (split.)^{35}S	87.1 days	β^-	0.167	^{35}Cl	1.4×10^{-3}	4.5
^{36}Cl	$^{40}Ar(p, 2p3n)^{36}Cl$	3.1×10^{5} years	β^-	0.714	^{36}Ar	1.1×10^{-3}	1.5×10^{6}
^{39}Ar	$^{40}Ar(n, 2n)^{39}Ar$	270 years	β^-	0.565	^{39}K	5.6×10^{-3}	
	$^{39}K(n, p)^{39}Ar$						
	$^{38}Ar(n, \gamma)^{39}Ar$						
^{53}Mn	$^{53}Fe(p, 2p)^{53}Mn$	2×10^{6} years	K-capture		^{53}Cr	$< 10^{-7}$	
	$^{56}Fe(p, \alpha)^{53}Mn$						
^{59}Ni	$^{59}Co(p, n)^{59}Ni$	8×10^{4} years	K-capture		^{59}Co	$< 10^{-7}$	
	$^{60}Ni(p, pn)^{59}Ni$						
^{81}Kr	$^{82}Kr(n, 2n)^{81}Kr$						
	$^{80}Kr(n, \gamma)^{81}K$	8.1×10^{-5} years	K-capture		^{81}Br	1.5×10^{-7}-10^{-5}	

Table 7.3 Distribution of Cosmogenic Radionuclides in the Principal Exchange Reservoir [7.3, 7.9]

Radio-nuclides	Exchange Reservoir (ppm)						Specific Activities		
	Stratosphere (S)	Troposphere (T)	Continental reservoir (B + G)	Mixed oceanic layer (M)	Deep oceanic layer (D)	Oceanic sediment (P)	dpm/kg air		dpm/ton of oceanic water
							stratosphere	troposphere	
^{10}Be	3.7×10^{-7}	2.3×10^{-3}	0.29	5.7×10^{-6}	10^{-4}	0.71	—	—	10^{-3}
^{26}Al	1.3×10^{-6}	7.7×10^{-8}	0.29	1.4×10^{-5}	7×10^{-5}	0.71	—	—	1.2×10^{-5}
^{36}Cl	10^{-6}	6×10^{-8}	0.29	1.4×10^{-2}	0.69	0	—	—	0.55
^{14}C	3×10^{-3}	1.6×10^{-2}	4×10^{-2}	2.2×10^{-2}	0.92	4×10^{-3}	—	—	260
^{32}Si	1.9×10^{-3}	1.1×10^{-4}	0.29	3.5×10^{-3}	0.68	2.8×10^{-2}	—	—	2.4×10^{-2}
^{39}Ar	0.16	0.83	0	2×10^{-4}	3×10^{-3}	0	—	—	2.9×10^{-3}
^{3}H	6.8×10^{-2}	4×10^{-3}	0.27	0.35	0.3	0	6	7×10^{-2}	36
^{22}Na	0.25	1.7×10^{-2}	0.21	0.44	8×10^{-2}	0	5×10^{-3}	6.7×10^{-5}	—
^{35}S	0.57	8×10^{-2}	0.10	0.24	4×10^{-3}	0	0.28	7.8×10^{-3}	—
^{7}Be	0.60	0.11	0.08	0.2	2×10^{-3}	0	0.17	0.63	—
^{33}P	0.64	0.16	5.6×10^{-2}	0.13	7×10^{-4}	0	0.15	7.6×10^{-3}	—
^{32}P	0.60	0.24	4.7×10^{-2}	0.11	10^{-4}	0	0.17	1.4×10^{-2}	—

Cosmogenic $^{14}CO_2$, and Ar and Kr isotopes leave the atmosphere at a much slower rate via molecular exchange at atmosphere-hydrosphere interface. The rate of formation of cosmogenic radionuclides and their distribution between the main reservoirs of the Earth can only be derived from the figures obtained prior to nuclear tests in the atmosphere (Table 7.3 and Fig. 7.2). Nuclear explosions markedly increased the concentration of radionuclides in the atmosphere.

7.3 Cosmogenic Radionuclides as Tracers of Natural Processes

Cosmogenic radionuclides present infinitesimal portion of the overall number of radionuclides on Earth (Tables 7.2 and 7.3). All of them are relatively short-lived and are only present in terrestrial compounds due to nuclear reactions. Since prior to nuclear tests there were no other sources of them on earth, cosmogenic radionuclides can characterize processes in those reservoirs where they are accumulated most.

Table 7.3 shows that most of ^{10}Be, ^{26}Al, ^{32}Si accumulate in oceanic sediments. At the same time ^{3}H, ^{7}Be, ^{36}Cl, ^{22}Na, ^{35}S, and ^{39}Ar are absent. Oceanic water and the surface waters of the continents contain most of ^{3}H and ^{14}C. The radionuclide concentrations in underground waters depend on the exchange conditions with atmospheric waters. If the exchange with the atmospheric reservoir ceases, the cosmogenic radionuclides gradually decay. The time (t) that has passed since the object under investigation has been removed from exchange with the atmosphere can be assessed by the formula:

$$t = \frac{1}{\lambda} \ln \frac{N_0}{N_t} \tag{7.1}$$

where N_0 is the concentration of the cosmogenic isotope in the sum of natural isotopes for a complete exchange with the atmosphere, and N_t is the concentration in the sample.

The principal conditions for the use of cosmogenic radionuclides for establishing the time-dependent characteristics of geological processes are:

1. The intensity of cosmic radiation and, consequently, the concentration of radionuclide should be practically constant during the time covered by the corresponding method.

2. The time needed for the exchange of the radionuclide between the atmospheric and the terrestrial reservoirs should remain constant and considerably smaller than the nuclide half-life.

3. The system should be closed relative to the investigated radionuclide after the sample leaves the exchange cycle, i.e. the isotopic composition should change only as a result of radioactive decay.

In selecting the radionuclide as a tracer it is necessary to take into account that its half-life should be commensurable with the time scale of the exchange of the process being investigated. For instance, the average time for deep oceanic water circulation is of the order of 1000 years. Hence, to study the circulation of water in the ocean we can use ^{14}C ($T_{1/2}$ = 5730 years) rather than ^{36}Cl (300 000 years), since it is too long-lived or ^{3}H because it is too short-lived as compared with the time of circulation.

We must also take into consideration the chemical compounds which include cosmogenic radionuclides. For example, ^{3}H, ^{36}Cl, ^{14}C are found in sea water in the form of soluble compounds (e.g. $^{1}H^{3}HO$, ^{36}Cl, $^{14}CO_2$). It is precisely these radionuclides that can serve as tracers for water, or the salts that precipitate (carbonates, chlorides). ^{32}Si and ^{10}Be come with aerosol particles and then with a suspension that in sea water precipitate to the bottom. Therefore, ^{32}Si and ^{10}Be are better tracers of accumulation of deep-water sediments.

Cosmogenic radionuclides have been found useful when studying:

1. the rate of sedimentation and the geochronology of oceanic sediments (^{10}Be, ^{26}Al, ^{32}Si, ^{14}C);
2. the circulation and mixing of oceanic water (^{3}H, ^{14}C, ^{36}Cl, ^{39}Ar, ^{7}Be, ^{32}Si);
3. the age and dynamics of continental underground waters (^{3}H, ^{14}C, ^{36}Cl);
4. the genesis of the recent evaporites (^{36}Cl);
5. the rates of weathering, erosion and sedimentation on continents (^{10}Be, ^{26}Al);
6. tephrochronology (^{14}C);
7. the age of marine terraces (^{14}C);
8. the age of continental Quaternary deposits and soils (^{14}C);
9. the process of subduction (^{10}Be);
10. archaeology (^{14}C);
11. meteorology: the circulation and mixing in the stratosphere and the atmosphere, the exchange between the atmosphere, hydrosphere, etc. (^{22}Na, ^{35}S, ^{7}Be, ^{37}Ar, ^{33}P, ^{32}P, ^{3}H, ^{14}C, etc.);
12. the age and time of the movement of glacial ice (^{32}Si, antropogenic ^{137}Cs).

7.3.1 Radiocarbon Method of Dating

Radiocarbon dating is one of the most reliable geochronological methods. It relies on the natural decay of ^{14}C, which comes into the biosphere or hydrosphere from the atmosphere. The hydrosphere is enriched with ^{14}C

as a result of the molecular exchange by CO_2 between the atmospheric and surface water, which takes place at the rate of about 20 mol/m^2 per year. The absorption of ^{14}C by the biosphere occurs due to photosynthesis in plants and the subsequent uptake by animals. Radiocarbon enters the carbonaceous skeletons and shells of sea organisms (as is the case with hydrobionts in general) from the water.

The average content of cosmogenic ^{14}C in the exchange carbon of the biosphere is $1.07 \times 10^{-10}\%$, while the specific activity varies from 13 to 17 decays per gram of carbon per minute. When carbon exchange is stopped (the death of the organisms or the formation of carbonates) the concentration of radiocarbon in these samples begins to diminish because of decay, which is not compensated, and in the course of time falls down to zero.

These formations can be used for radiocarbon dating. The upper limit of the method is determined by the half-life of ^{14}C (5730 years) and the sensitivity of the analysis. After approximately 60 thousand years, 0.1% of the original radiocarbon remains which determines the upper limit of the method. The lower limit is restricted by the accuracy of the analysis. With a 0.3% minimal error it is 80 years.

In principle, any carbon-containing formations no older than 60 thousand years may be an object of radiocarbon dating. The best are wood remains, charcoal, and shells. Since one of the conditions for the reliability of any geochronological method is that the system is closed, then, when analysing the ^{14}C in samples, it is always essential that we should take into account the possibility of contamination by recent (or ancient) carbon brought by ground water in the form of organic acids, hydrocarbonates, and carbonates. It should also be mentioned that various methods for cleansing the surfaces of samples are employed.

The concentration of ^{14}C is determined by means of comparing it with the standard containing recent (or ancient) carbon

$$\delta^{14}C = \frac{\left({}^{14}C/{}^{12}C\right)_{sam} - 0.95\left({}^{14}C/{}^{12}C\right)_{stand}}{0.95\left({}^{14}C/{}^{12}C\right)_{stand}} \times 1000[‰] \tag{7.2}$$

where ‘stand’ means ‘standard’ and ‘sam’ stands for ‘sample’. The coefficient 0.95 covers dilution by non-radioactive ancient carbon. As an international standard the oxalic acid of the National Bureau of Standards (USA) is used. At present the NBS standard has an activity of 13.56 decays/min.

Chemogenic carbon in carbonates is approximately by 5% more active than the carbon of the same age but of vegetative origin. This is due to the fractionation of carbon isotopes during photosynthesis. The lighter isotopes are absorbed by plants more readily than the heavier ones.

The factor which promotes radiocarbon application is the short time of the exchange between the atmospheric and terrestrial reservoirs (about 10 years) as compared with the half-life. Another important condition is that the isotopic composition of carbon in the atmosphere over the past 100 thousand years remains constant, which is explained by a constant neutron flux and perfect mixing of the CO_2 in the atmosphere. The latter fact can be estimated from the same concentrations of ^{14}C (within 3-5%) in objects of the same type in different parts of the world, irrespective of the fact that the neutron flux rises 3.5 times from the equator to the poles.

The variations in neutron flux over time have been studied by comparing the age of sequoia determined by the ^{14}C method with the number of its annual rings, or with the data of wood archeological remains. It was established that the neutron flux has varied with an amplitude of 1-3% and a period of 100-1000 years and with an amplitude of up to 8.5% and a period from 2200 to 6000 years.

These variations could have been brought about by astrophysical reasons (changes in the Earth's magnetic field, supernova outbursts), as well as by climate, which disturbs the C equilibrium in the main exchange reservoirs.

The activity of recent atmospheric carbon has greatly changed due to anthropogenic factors. Over the past 100 years atmospheric carbon dioxide has been diluted by non-radioactive ancient carbon formed by burning fossil fuel (the Suess effect). The concentration ^{14}C in the atmospheric carbon dioxide had decreased by 2-5% by the middle of the century. Beginning from the 1950's, ^{14}C content increased in the atmosphere due to nuclear tests and was 2.5 times more at the end of the sixties. After the termination of nuclear tests in the atmosphere the concentration of ^{14}C gradually decreased and at present exceeds the natural concentration by approximately 30%. This decrease has been due to the absorption of ^{14}C by the ocean.

Radiocarbon dating has been extensively applied in geology, hydrogeology, oceanology, and archeology to date events in the Upper Quaternary period.

The following examples illustrate its use.

1. The study of the isotopic composition of sulphur in the bottom silts of the Black Sea agrees well with geological and biological observations and has shown that the hydrosulphuric contamination of the sea began after seawater broke through the Bosporus to transform the fresh-water neo-Euxine Sea into a salt basin. In water rich in sulphates and in the absence of adequate transverse mixing the biogenic reduction of sulphates began. It became possible to establish the time of the emergence of hydrosulphuric contamination by applying the radiocarbon method of investigating the organic substance of silts. Vinogradov and Devirtz discovered that about 7500-8000 years ago the saline waters of the Mediterranean Sea began to fill the ancient Black Sea basin, the process lasting nearly one hundred years.

2. Radiocarbon dating is employed to investigate Quaternary volcanism. The analysis of charred trees captured by the lava during the eruption makes it possible to determine the age of lava streams. Tephrochronology has now gained acceptance. The age of soils and wood remains buried under the layers of volcanic ash (tephras) are established by the method.

3. The age of soils is estimated through a ^{14}C analysis of soil humus. The humus of podzol soils at 70-80 cm depth ages 2.5 thousand years. The formation of black soils is more durable and for a thickness of 90-100 cm lasts 5-5.5 thousand years [7.9].

4. The dating of marine terraces by the mollusk shells yields the age of marine transgressions and regressions which led to the formation of these terraces. According to Leontyev, the latest transgressions of the Caspian Sea occurred within the time interval of 24-22 centuries B.C., 10-9 centuries A.D., and 11-9 centuries A.D. The sea level has risen by 20-40 m [7.9].

5. ^{14}C is comprehensively used to establish the age of underground water. It is assumed that near the surface water has a constant equilibrium concentration of ^{14}C. The concentration falls in compliance with the law of radioactive decay as soon as the water penetrates into the deeper layers of the Earth's crust.

The age of water can be derived from the equation (7.1):

$$t = \frac{1}{\lambda_{14}} \ln \frac{^{14}C_0}{^{14}C_t}$$

However, the obtained value will contain indeterminacy, since the way the carbonate system change during interaction of underground waters with water-enclosing rocks is not known.

It is possible to use the ratio of stable isotopes $^{13}C/^{12}C$ to trace the variation in the carbonate system in the water and introduce a correction for the exchange between the carbonate of aqueous solution and the carbonates of water-enclosing rocks. For example:

$$\delta^{14}C = \frac{\delta^{14}Co\delta^{13}C_{HCO_3^-}}{\delta^{13}C_{CO_2}} - e^{-\lambda_{14}t} \tag{7.3}$$

where

$$\delta^{13}C = \left[\frac{\left(^{13}C/^{12}C\right)_{sam} - \left(^{13}C/^{12}C\right)_{stand}}{\left(^{13}C/^{12}C\right)_{stand}} \right] \cdot 100\%$$

7.3.2 Tritium Method of Dating

This method uses the radioactive decay of tritium, which is continuously formed in the upper atmosphere and falls to the surface of the Earth with precipitation in the form $^1H^3HO$. The half-life of tritium is 12.3 years. Therefore the method is applied to the study and date young natural waters associated with atmospheric precipitation, whose age is less than 60-70 years.

The only source of tritium is atmospheric water. When tritium gets into water of the difficult water-exchange, it gradually decays and practically disappears within 60-70 years. The presence of tritium in underground water unambiguously signifies it has atmospheric water as its source. The time that has passed from the precipitation of atmospheric water can be found from formula 7.1. Tritium is mainly formed in the atmosphere by the reaction $^{14}N(n, t)^{12}C$ with $E_n > 4.4$ MeV.

It has been estimated that only 3-5% of the secondary neutrons formed in the atmosphere by cosmic radiation form tritium. According to Suess, the rate at which cosmogenic tritium is formed is approximately 30 atoms per cm²/min. The concentration of natural tritium is very low and comprises about 10^{-18} of atmospheric hydrogen. Hence, the concentration of tritium is usually expressed in special 'tritium' units (TU)

$$1\ \text{TU} = 10^{-18}\ \frac{^3H}{^1H} = 7.2\ \text{decay/min} = 3\ \text{pCi per litre of water}$$

Prior to nuclear tests the concentration of tritium varied from 0.1 to 10 TU, while the overall quantity on Earth was 34 MCi. Most of this (~93%) was in the hydrosphere and only 7% in the atmosphere. From 1952, when nuclear tests began, 'bomb' tritium appeared in the atmosphere. Most of it was formed by the nuclear fusion reaction:

$$^2_1H + ^2_1H \rightarrow ^3_1H + ^1_1H + 4.0\ \text{MeV}$$

The first nuclear explosions were not powerful and carried out at a relatively small altitude. Hence, all the bomb tritium was concentrated in the troposphere and was then rapidly transferred with precipitates, into the ocean. Beginning from 1958 high altitude explosions with a power of several megaton were started and the intensity of the tests reached a maximum in 1961-62. As a result, a large amount of tritium moved into the stratosphere, where its total amount reached 200 kg (the overall total of cosmogenic tritium being 1.8 kg).

At that time the ^3H concentration in atmospheric precipitation was as high as 10 000 TU in different parts of the Earth. Most of the bomb tritium has now disintegrated and was estimated in the seventies to be 4500 MCi.

Tritium dating is applied to study young contemporary waters and to find the connection between ancient and atmospheric waters.

The following examples are illustrative.

1. The investigation of tritium in artesian basins makes it possible to distinguish the alimentation areas of the basin by water. Artesian water of Paleozoic or Mesozoic age cannot preserve primary atmospheric water tritium. The presence of noticeable concentrations of tritium reveals a connection between ancient waters and atmospheric waters.

2. The $^3H/^1H$ ratio in ocean water can establish the age of water layers at different depths and thus determine the rate at which they mix. Thus, it has been discovered that the upper 50-meter layer of the Atlantic Ocean is gradually mixed within 1-2 years. The water in intermediate layers (60-180 m) was almost fifty years old. The water in very deep layers contains infinitesimal amounts of ^3H and by ^{14}C dating is about 600 years old. Large concentrations of tritium were unexpectedly found in bottom waters of the Atlantic Ocean. It was assumed that some surface layers at high latitudes rapidly lowered and moved southward without mixing with the upper layers.

3. Tritium is widely used in the quantitative estimate of the exchange of natural water. To achieve this goal various models are used. For instance, to determine the time of exchange within individual underground drainage systems, represented by springs, flowing at different heights, the model of a perfectly mixed reservoir is applied. The model presumes the volume of the reservoir to be constant, and the concentration of tritium in it to be homogeneous and equal to that at the exit from the system. The tritium is supplied at the entrance by atmospheric precipitation. With this in view, the following relation [7.9] is found to be adequate:

$$C_n(\text{exit}) = C_{n-1}(\text{exit})\; e^{(\varepsilon+\lambda)} + C_n(\text{entrance})\; \frac{\varepsilon}{\varepsilon+\lambda}\left[1 - e^{(-\varepsilon+\lambda)}\right]$$

where C_n (exit) is the concentration of tritium in underground waters at 'n' years; C_{n-1}(exit) is the concentration of tritium at (n – 1) years; C_n (entrance) is the concentration of tritium in atmospheric precipitation at n years; λ is decay constant for tritium (λ_3 = 0.055 year^{-1}); ε is a dimensionless constant equal to the ratio of the annual precipitation to the volume of the basin (1/ε is the period of basin water exchange). In this case, the difference is used between the concentration of tritium prior to the thermonuclear tests and after them. This difference has led to the emergence of peaks in the tritium concentration in underground waters, which are considered as specific temporal datum marks.

7.3.3 Beryllium Method of Dating

The beryllium method is based on the decay of the ^{10}Be beryllium isotope which is formed under the action of cosmic rays and is adsorbed by particles of dust, and falls to the Earth's surface with the rain [7.10]. The half-life of ^{10}Be is 1.5 mln years. Thus, the beryllium method can be employed for the time interval of 10 mln years, viz. between the lower limit for argon method and the upper limit of the ionium method.

If we assume that over the past 10 mln years the rate of sedimentation and that of ^{10}Be supply to a sediment remained constant, we would expect a smooth decrease in its concentration with depth.

Hence, we may use beryllium dating to determine the age of sediments and the rate at which they are deposited. Besides, as with other methods, the vertical migration of ^{10}Be should be absent.

The insignificant influence of vertical migration is supported by a smooth drop in activity with depth, as well as by the constancy of the concentration of the stable ^{9}Be. The rate of sedimentation can be determined according to the formula [7.11]:

$$C_x(t) = \frac{p}{s}\exp(-\lambda x / s) \tag{7.4}$$

where $C_x(t)$ is the concentration of ^{10}Be in the layer of sediments at a depth of 'x' cm from the surface which was sedimented 't' years ago; p is the rate of accumulation of ^{10}Be atoms by the sediment (1.8×10^{-12} atom/cm^2·s); s is the rate of sedimentation (mm/10^3 years); and λ is the ^{10}Be decay constant (4.33×10^{-7} years^{-1}).

For $x/s << 1$ the equation is simplified:

$$C_x(t) = \frac{p}{s} \tag{7.5}$$

2. To calculate the rate of the growth of ferromanganese concretions, the distribution of ^{10}Be with depth is used (Somayajulu, 1967):

$$C^* = \frac{R^*}{\lambda\rho(X_2 - X_1)} \exp\frac{-X_1 \cdot \lambda}{s} - \exp\frac{-X_2}{s}$$

where C^* is the concentration of ^{10}Be;

R^* is the rate of ^{10}Be accumulation in the concretion;

ρ is the density of the concretion;

λ is the ^{10}Be decay constant;

s is the rate of the growth of the concretion (mm/10^3 years);

X_1 and X_2 are the depths of the layers of concretion ($X_0 = 0$ at the surface).

If we assume that the rate of concretion growth and that of accumulation of ^{10}Be and ^9Be were constant, then

$$C = R/s\rho$$

The rate of sedimentation calculated from ^{10}Be is consistent with the values determined by other methods. Moreover, the rate of ^{10}Be sedimentation in silts equal to 2.55×10^6 atm/cm^2 per year is consistent with the rate obtained by investigations of nuclear reactions. These methods established the constancy of the formation rate of ^{10}Be in the atmosphere, and, consequently, the constancy of radiation within the past 10 million years. On the basis of chronological research of oceanic sediments it was discovered that the sedimentation rate of ^{10}Be from the atmosphere depends on latitude: in equatorial and middle latitudes it is higher (1-3 atom/cm$^2\cdot$s) than in polar latitudes (0.6 atom/cm$^2\cdot$s).

3. Another method has been suggested to determine the rate of erosion, weathering, or sedimentation on the continent using ^{10}Be and ^{26}Al produced in quartz by cosmic rays according to reactions with Si and O. Since the concentration of Be and Al in quartz is low, the isotopic ratios $^{26}Al/^{27}Al$ and $^{10}Be/^9Be$ can be high [7.12].

4. ^{10}Be as a geochemical tracer was used to confirm the capture of oceanic sediments during subduction [7.13]. It was assumed that if this phenomenon actually took place and the time from the sediment subduction to the formation of island arcs and volcanic rocks did not exceed 10 mln years, these rocks should contain measurable amounts of cosmogenic ^{10}Be. And indeed, in recent lavas of young island arcs considerable quantities of ^{10}Be were discovered (Kuril island arcs—3.1×10^6-8.0×10^6 atom/g; Aleutian Island arcs—3.8×10^6 atom/g). The concentration of ^{10}Be on the surface of oceanic sediments is 5×10^6 atom/g. In more ancient arcs (Japan, Mariana, Sunda, etc.) the concentration of ^{10}Be was lower than the detection threshold (10^6 atom/g). In recent volcanic rocks of different origin (riftogenic, island) no ^{10}Be has been found.

7.3.4 Other Cosmogenic Radionuclides

The ^{32}Si, ^{36}Cl, and ^{10}Be isotopes have not yet been extensively applied due to their extremely low concentrations and activities and the analytical difficulties. However, there are some papers which show that cosmogenic radionuclides may well be fairly promising geochemical tracers.

1. ^{32}Si has been successfully used to determine the sedimentation rate of oceanic sediments. ^{32}Si is accumulated on the surface of sediments with particles of dust precipitated from the atmosphere. The fraction of ^{32}Si in natural silicon is 10^{-17}. The evaluation method is analogous to the assessments of other cosmogenic radionuclides [7.14]. ^{26}Al can be analogously used [7.15].

2. Cosmogenic ^{36}Cl can serve as a tracer in sea water. Its concentration in recent water (^{36}Cl/^{35}Cl) is $(1.7\text{-}2.0) \times 10^{-4}$.

For example, ^{36}Cl was employed to study the genesis of evaporites at the bottom of the Dead Sea [7.16]. It was shown that the evaporites were not formed during recent evaporation of sea water, but were formed from ancient saline diapirs below the Dead Sea itself that does not contain cosmogenic ^{36}Cl.

7.4 Nuclear Reactions in the Lithosphere

The main driving force of nuclear reactions in the lithosphere is radiation of natural radionuclides and secondary neutrons formed by these reactions. The role of cosmic radiation in initiation of nuclear reactions is insignificant, since most of it is absorbed by the topmost layer of the earth's crust. The rapid absorption and scattering of natural radiation in rocks leads to local nuclear reactions. The radionuclides thus formed are present in minute concentrations and have no geochemical significance.

Nuclear reactions in the lithosphere are caused by α-particles, γ-photons and neutrons. The energy of β-particles formed in natural β-decay is too small to induce nuclear reactions.

A. The sources of γ-radiation in the lithosphere are natural radionuclides. The energies of natural γ-rays do not exceed 3 MeV, which is considerably less than the bond energy of nucleons in the nucleus for most elements. Consequently, γ-radiation in nature can only induce (γ, *n*) reactions in deuterium and beryllium, i.e.,

$$^{2}_{1}\mathrm{H} + \gamma \rightarrow {}^{1}_{1}\mathrm{H} + n; \quad {}^{9}_{4}\mathrm{Be} + \gamma \rightarrow {}^{8}_{4}\mathrm{Be} + n$$

The beryllium reaction results in the unstable ^{8}Be isotope of beryllium and a neutron. Beryllium-8 almost instantly ($T_{1/2} = 10^{-15}$ s) decays into two α-particles:

$$\mathrm{Be} \rightarrow 2\,^{4}_{2}\mathrm{He}$$

Thus, the two reactions are a natural source of neutrons, and the second yields an α-particle.

B. Natural α-particles induce (α, *n*) and (α, *p*) reactions and can penetrate the nuclei lighter than potassium (viz., Li, Be, B, N, O, F, Na, Mg, Al, C, S, Cl). Given the small penetrability of alpha particles, we may assume that such reactions will only occur when the atoms that are the sources of radiation and the target atoms are concentrated in the same mineral. The reaction (α, *n*) is most intense with beryllium:

$$^{9}_{4}Be + ^{4}_{2}He \rightarrow ^{12}_{6}C + n$$

The reaction is an important source of neutrons. Radioactive minerals containing beryllium impurities, such as tscheffkinite, orthite, cyrtolite, orangite, and niobotantalites, are natural neutron sources.

C. The most intense nuclear reactions taking place in the lithosphere are those induced by neutrons. Neutron sources in the lithosphere include the (α, *n*) and (γ, *n*) reactions, the spontaneous fission of uranium, and cosmic neutrons. The principal source of neutrons is the (α, *n*) reactions. It yields over 80% of the neutrons found in the Earth's matter.

The neutron flux varies in different rocks. Measurements have shown that in schists and marbles it is 5 neutron/day·cm^2; in granites-rapakivi it is 13 neutrons/day·cm^2, in rocks with beryllium deposits it is 1500 neutrons/day·cm^2. In uranium minerals more than half neutrons (up to 90%) originate from spontaneous fission. On the average a single fission event yields about for two neutrons (2.6). Most spontaneous fission events come from ^{238}U, since its half-life is considerably smaller than that of ^{232}Th or ^{235}U ($t_{1/2}$ for $^{232}Th \geq 10^{22}$ years, $t_{1/2}$ for ^{235}U is 1.8×10^{18}, and $t_{1/2}$ for ^{238}U is 8.7×10^{17} years (Shukolyukov, 1982)). The nuclear reactions induced by neutrons depend on the neutron energy. Both fast and slow neutrons have been discovered in rocks.

The neutrons formed in rocks due to (α, *n*) and (γ, *n*) reactions have relatively low energies. The main type of reaction with nuclei is (*n*, γ) type typical for nuclei with large effective cross sections (e.g., Cd, B, Cl, Mn, the rare earths). Some lighter elements (e.g., Li, B) undergo (*n*, α) reactions, e.g.

$$^{6}_{3}Li + n \rightarrow {}^{3}H + \alpha$$

The rate at which tritium is formed in the lithosphere is 10^{-3} atom/cm^2·s. However, in pore water in granites with high concentrations of uranium and thorium (neutron sources), and of Li (a target nucleus) and a low concentration of B (a neutron absorber), the content of tritium can reach 2.5 T.U. Assessing the age of waters in this case by tritium dating will be inaccurate (Andrews, Kay, 1982).

Moreover, thermal neutrons can bring about the fission of ^{235}U nuclei. This process involves the release of additional neutrons, the contribution of which to the overall neutron flux in uranium deposits can occasionally become as high as 50%.

Fast neutrons are either of cosmic origin or are due to the fission of heavy nuclei and can react with other heavy nuclei. These reactions lead to ^{239}Pu in uranium minerals, i.e., $^{238}U + n\ ^{239}U \xrightarrow{\beta} {}^{239}Np \xrightarrow{\beta} {}^{239}Pu \xrightarrow{\alpha} {}^{235}U$.

As a result of the spontaneous fission of ^{235}U radioactive fragments accumulate in rocks. The quantity of such fragments in the lithosphere is quite insignificant. Fragments give rise to the chains of radionuclides. Each of the 70 radioactive chains which emerge in the course of fission end with a stable nuclide. The concentrations of isotopes formed in this way are 5-9 orders of magnitude lower than those of the same isotopes that became a part of the Earth since it formed. The exceptions are the isotopes of krypton and xenon formed in spontaneous fission whose number is only by 2-3 orders lower than their original concentration.

7.5 The Spontaneous Chain Reactions in Uranium Deposits

Nuclear fission triggered by neutrons can occur in isotopes of uranium and thorium. To induce fission in the most widespread isotopes ^{238}U and ^{232}Th, the energy of the neutrons should exceed 1 MeV. High energy neutrons are either of cosmogenic origin, or appear during spontaneous fission. However, as a result of inelastic scattering by the nuclei of petrogenic elements they rapidly give up their energy, and are decelerated to energies that are insufficient for a fission reaction. The only natural radionuclide capable of a fission reaction with neutrons of various energies is ^{235}U. Its concentration today is low, being 0.72% in mixture of natural uranium isotopes. With such low concentration, ^{235}U does not reach critical mass even in rich uranium ores. At the early stages of the Earth's development the ^{235}U concentration was much greater (Table 7.4).

The size an ore body needs to be for a chain reaction to develop depends on the age of the ore and the concentration of uranium in it. In Phanerozoic deposits no chain reaction is possible irrespective of the dimensions of the body.

According to the data [7.17] for ores older than 1.5×10^9 years and whose concentration of uranium is also about 50%, a chain reaction could have taken place even in ore bodies of insignificant thickness (0.5-1.0 m^3). In the case of depleted ores the volume must be much larger. Thus, for ores 2.5 bln years old, a halving of the concentration of uranium requires a ten times greater

Table 7.4 Change in Isotopic Composition of Uranium during the History of the Earth [7.17]

Time up to the present day, bln years	0	0.5	1.0	2.0	3.0	4.0	4.6
$^{238}U/^{235}U$	137.7	91.6	60.8	26.85	11.85	5.20	3.23
Fraction of ^{235}U in the mixture of isotopes, %	0.72	1.08	1.62	3.59	7.8	16.1	23.6

volume. The cross-section of the fission reaction of ^{235}U is the largest for thermal neutrons. Therefore, an important function in a natural nuclear reactor would be played by neutron moderators (H, C, O), which are concentrated in water and organic matter. The concentration of these elements has an important role in the initiation of a chain reaction and affects the critical mass. For instance, for an ore body 2×10^9 years old and with a uranium concentration of 50% and a water concentration of 20%, the critical volume is 0.6 m^3, while for an 18% water content it is about 6000 m^3.

The Natural Nuclear Reactor in Oklo

The only natural nuclear reactor known on Earth is the uranium deposit in Oklo (Gabon, West Equatorial Africa). The reactor was discovered by French scientists in 1972 after finding a systematic lack of ^{235}U in the ores, which is due to the burning out of fissioned radionuclide. The $^{235}U/^{238}U$ ratio was 0.6% on average, reaching 0.3% in some of the samples (the normal ratio being 0.72%). No less than 500 tonnes of highly depleted uranium were found in this deposit, which corresponded to the release of about 10 000 MW of energy a year. The deposit is sedimentary-epigenetic in origin and is bedded in terrigenous mid-Precambrian rocks. The source of the uranium was the Archean rocks of the crystalline foundation, in which the uranium was mobilized during erosion. The precipitation of the element is associated with the reducing barriers in aluvial formations. Later diagenetic, and particularly epigenetic transformations led to a high uranium concentration and the formation of the present-day ore body.

The average uranium concentration in the ore is 0.4-0.5%. The uranium is in the composition of the carbon-bearing cement in coarse-grained sandstones. The deposit has four ore zones with a uranium concentration of up to 50%. Each zone is 20-50 m from each other. The size of the zone is nearly 5×30 m, with a volume of 300 m^3. These zones are correlated with argillite lenses in terrigenous sandstones. It is these zones that form the natural reactor in Gabon. Thus, this deposit had a whole series of natural nuclear reactors, which affected each other with their neutron fields.

The radiological age of the deposit was assessed by various methods, yielding an approximate value of 1.8 bln years. According to the estimates, the reactor functioned for several hundred thousand years (200-600 thousand years). Geochemical and mineralogical investigations indicated that the temperature in those active zones was 400-600 °C.

A long lasting chain reaction should lead to isotopic shifts due to the accumulation of the decay products of the fission fragments and to the nuclear reactions, predominantly induced by neutrons. A large number of elements in the ore deposits are distinguished by anomalous isotopic composition [7.17]. The elements that are particularly sensitive to a neutral flux are the rare-earth. Their isotopic compositions have a marked heterogeneity and anomaly within the confines of the ancient reactor.

The heterogeneity reflects the diverse origin of each of the isotopes: viz., the primary natural distribution is changed because of the accumulation of fission products or generation, as well as the burning out as a result of neutron reactions. For instance, gadolinium isotopes in a natural mixture make up 0.2% for ^{152}Gd and 2% for ^{154}Gd. In the reactor deposit the isotopic concentrations are 30% for ^{152}Gd and 20-40% for ^{154}Gd. Fragments from the fission reactions of ^{235}U, ^{238}U and ^{239}Pu, which could have emerged by the bombardment of ^{238}U by fast neutrons, do not contain ^{152}Gd and ^{154}Gd. Presumably they emerged from europium isotopes as a result of (n, γ) reactions, viz., $^{151}\mathrm{Eu}(n, \gamma)^{152}\mathrm{Eu} \xrightarrow{k} {}^{152}\mathrm{Gd}$ and $^{153}\mathrm{Eu}(n, \gamma)\,^{154}\mathrm{Eu} \xrightarrow{k} {}^{154}\mathrm{Gd}$. As a result of this the isotopic distribution of ^{151}Eu and ^{153}Eu should be smaller, as is the case in the Gabon ores.

The isotopic compositions of Xe and Kr in ores were also distinct from the natural ones and were close to the fission products of ^{235}U with their thermal neutrons. Chain reactions in these deposits led to a change in the isotopic composition of argon, thus increasing the concentrations of ^{36}Ar and ^{38}Ar. Isotopic shifts can also be found in the composition of other less extensively distributed elements. The fraction of original non-radiogenic Ru in the ores of these deposits is less than 10% and that of Pd is 4%. The anomalous isotopic composition of cadmium reflects a large cross-section of the (n, γ) reaction for the ^{113}Cd isotope. The burning of ^{113}Cd as a result of the $^{113}\mathrm{Cd}(n, \gamma)^{114}\mathrm{C}$

reaction results in a drop in the concentration of the isotope from 17.75% to 1.25% and an increase in the fraction of ^{114}Cd from 17.75 to 29.60-31.79%.

The anomalous isotopic compositions of a large number of elements have made it possible to reconstruct the neutron flux in the Gabon deposit whilst the reactor was operating and was estimated to be close to 10^{21} neutron/cm^2·s. The fraction of neutrons with energies more than 1 MeV was 20-50% of the number of thermal neutrons.

The secondary migration of elements from the reactor could lead to the depletion of the reactor zone of radiogenic isotopes, and to a distortion in the isotopic composition as a result of isotopic exchange. Research indicates that the migration of Xe and Kr was intense, though no exchange with non-radiogenic Xe and Kr was observed.

The migration of alkali and alkali-earth elements (Rb, Sr, K, Ba, Li) was accompanied by the depletion of radiogenic isotopes and the influx of non-radiogenic ones, which led to the establishment of the average isotopic compositions and their approximation to the normal spectrum. The least mobility was exhibited by the rare-earth elements. The isotopic composition of Nd remained unchanged. Under the above conditions, uranium also migrated poorly which is substantiated by the preservation of the ore and the isotopic composition of uranium in uranium minerals.

The investigations show that the conditions favourable for a natural chain reaction of uranium fission could only have existed for a relatively short interval of time from 2.6 to 1.5 bln years. Earlier, the geochemical circumstances were unfavourable for the formation of a sufficiently rich uranium ores. Later, the decomposition of ^{235}U lowered its concentration to values insufficient for critical mass.

No other uranium deposits with traces of natural nuclear reactors have been found on Earth. Such reactors have been presumably destroyed by recent erosion, or the erosion has not yet proceeded sufficiently to detect them.

7.6 Nuclear Reactions in Meteorites

Against the background of very low concentrations of the products of natural nuclear reactions in the lithosphere, formations where the accumulation of cosmogenic radionuclides exceed those of the primary radionuclides can also be found on Earth. These structures are formed by meteorites, which have been permeated by cosmic particles as they wandered through outer space. Nuclear reactions are most distinctly expressed in small meteorites. Inelastic collisions between fast and super-fast cosmic particles and the atomic nuclei in the meteorites lead to two main types of reactions, viz. simple (p, xn) or (*xp*, *yn*) type nuclear reactions and the deep fission of atomic nuclei. This process is characterized by the formation of a wide range of product nuclei from close relatives to the large nucleus to protons and neutrons.

Nuclear reactions also result in the accumulation in the meteorite of a large quantity of stable cosmogenic products [7.18]. The total concentration of cosmogenic radionuclides in meteorites is estimated to be 10^{-7} g/g (whereas the concentration of U and Th is of the order of 10^{-8} g/g). The activity of individual radionuclides in meteorites that have recently fallen fluctuates from 0, *n* to 100*n* dec/min·kg.

The following measurements have been made in meteorites: ^{32}P ($T_{1/2}$ = 14.3 days), ^{51}Cr (27.5 days), ^{37}Ar (35.1 days), ^{58}Co (72 days), ^{56}Co (77.3 days), ^{46}Sc (84.2 days), ^{45}Ca (168 days), ^{57}Co (267 days), ^{54}Mn (300 days), ^{49}V (330 days), ^{55}Fe (2.60 years), ^{22}Na (2.62 years), ^{60}Co (5.26years), ^{3}H (12.6 years),^{44}Ti(~46 years), ^{39}Ar (269 years), ^{32}Si (~650 years), ^{14}C (5745 years), ^{59}Ni (1.10^5 years), ^{36}Cl (2.6×10^5 years), ^{26}Al (7.4×10^5 years), ^{53}Mn (3.7×10^6 years), ^{40}K (1.48×10^9 years), and ^{10}Be (1.5×10^6 years).

The accumulation of stable decay products from radioactive nuclides leads to an isotopic composition in meteorites that differs from that found on Earth. The isotopic ratios of noble gases in meteorites differ from the ones that exist in terrestrial conditions, viz., $^{20}Ne/^{22}Ne$, $^{36}Ar/^{38}Ar$ are ten times greater, $^{21}Ne/^{22}Ne$ is 30 times greater, and $^{3}He/^{4}He$ is $2.5 \cdot 10^5$ times greater. The ratios for the isotopes of Kr and Xe in the Earth's atmosphere $^{129}Xe/^{132}Xe$ is higher. The isotopic proportion of cosmogenic potassium is $^{39}K : {^{40}K} : {^{41}K}$ = 0.7 : 0.67 : 1 whereas that in terrestrial rocks is 13.5 : 1.7×10^{-3} : 1. In iron meteorites the isotopic ratio $^{40}Ti/^{48}Ti$ is 10 times higher than on Earth, while for $^{50}V/^{51}V$ exceeds less than 10 times. The presence of cosmogenic radionuclides in meteorites makes it possible to determine their terrestrial age, viz., the time that has elapsed since they fell. The zero-point is taken to be the concentration of cosmogenic radionuclides in recent meteorites with an analogous composition.

For stone meteorites ^{14}C is frequently used. In the case of iron meteorites the concentration of radioactive nuclides greatly depends on the size of the meteorite and the depth at which the sample was taken. Hence, to exclude any depth effect the age is calculated from the ratios of radioactive nuclides with different half-lives but the same excitation functions, e.g. $^{14}C/^{10}Be$, $^{44}Ti/^{26}Al$, and $^{26}Al/^{36}Cl$. Estimates of terrestrial ages of meteorites reach 1 mln years.

The 'radiogenic' age of a meteorite corresponds to the time when the parent body is destroyed, after which cosmogenic radionuclides accumulated. The age *t* can be approximated from the formula:

$$N = N_0 \, sFt \tag{7.6}$$

where N is the concentration of stable decay products of cosmogenic radionuclides, N_0 is the concentration of target nuclei, σ is the cross-section of the reactions (determined experimentally), and F is the intensity of cosmic radiation flux.

7.7 Nuclear Reactions as the Sources of Nucleosynthesis in the Universe

Nuclear reactions in stars are the source of all existing nuclides. At present it is believed that nucleosynthesis takes place under conditions that differ in time and P-T parameters [7.18, 7.19].

1. About 15×10^9 years ago, during the 'big bang', the cosmogenic synthesis of elements took place which continued for nearly one hour, leading to the synthesis of ^1H, ^4He, ^2H(?), ^3He(?).

2. Today, the nucleosynthesis also occurs during the evolution of stars. At a stage of the gas-dust cloud, under the gravitation, the initially cold rarefied matter gradually consolidates, warms up and a protostar is formed. The process of gravitational compression and heating of the star leads to nuclear synthesis in the heart of the star. As soon as energy released by the nuclear reactions begins to compensate for the losses due to radiation, the gravitational compression of the star ceases. This star, like our own Sun, will belong to the 'main sequence', where it can remain for many billion years. Proceeding from the composition of the Universe (~70% hydrogen, ~30% helium, and ~1% O, C, etc.), we can assume that the nuclear reactions in the heart of a star will be fusion reactions.

(a) Nuclear fusion begins with the combustion of hydrogen and its transformation into helium:

$$p + p \rightarrow d + e^+ + \nu_e$$

$$^3_2\text{He} + ^3_2\text{He} \rightarrow ^4_2\text{He} + 2p$$

$$^3_2\text{He} + ^4_2\text{He} \rightarrow ^7_4\text{Be} + \gamma$$

The reaction chain ends in one of the two ways depending on the temperature inside the star:

$$^7_4\text{Be} + e^- \rightarrow ^7_3\text{Li} + \nu_e; \quad ^7_3\text{Li} + p \rightarrow ^8_4\text{Be}, \quad \text{or}$$

$$^7_4\text{Be} + p \rightarrow ^8_5\text{B} + \nu_e; \quad ^8_5\text{B} \rightarrow ^8_4\text{Be} + e^+ + \nu_e, \quad ^8_4\text{Be} \rightarrow 2^4_2\text{He}$$

This chain of nuclear transformations is called the hydrogen (proton) cycle and takes place at (1-2) × 10^7 K. The result is the transformation of four protons into a ^{4}He nucleus with the release of 27.7 MeV of energy, most of which is carried away by neutrinos.

(b) When all the hydrogen burns up in the heart of the star, fusion only proceeds in the outer shell, which still contains hydrogen. The star expands and becomes a red giant, leaving the main sequence. Its fate depends on its mass. If its mass is less than 0.4 M_0 (the mass of the Sun), the evolution ceases when all the hydrogen is exhausted. The nucleus of more massive stars gradually contracts being deprived of any other source of energy. When the temperature in the heart of the star reaches (1-2) × 10^8 K, helium begins to burn (the alpha process):

$$3\,^4_2\text{He} \rightarrow \,^{12}_6\text{C} + 7.65\,\text{MeV}$$

When sufficient carbon is formed, helium will also start burning by the reaction:

$$^{12}_6\text{He} + \,^4_2\text{He} \rightarrow \,^{16}_8\text{O} + \nu$$

The alpha process lasts for about 10^4-10^6 years. This results in the accumulation of $^{16}_8$O and $^{12}_6$C.

The further combustion of carbon and oxygen calls for higher temperatures (> 5 × 10^8 K) and will only occur in very massive stars.

The process leads to the formation of ^{20}Ne, ^{23}Na, ^{24}Mg, ^{28}Si, ^{31}P, ^{32}S, thus,

$$^{12}_6\text{C} + \,^{12}_6\text{C} \rightarrow \,^{20}_{10}\text{Ne} + \,^4_2\text{He}$$

$$^{12}_6\text{C} + \,^{12}_6\text{C} \rightarrow \,^{23}_{11}\text{Na} + p$$

$$^{12}_6\text{C} + \,^{12}_6\text{C} \rightarrow \,^{23}_{12}\text{Mg} + n$$

$$^{12}_6\text{C} + \,^{12}_6\text{C} \rightarrow \,^{24}_{12}\text{Mg} + \gamma$$

$$^{16}_8\text{O} + \,^{16}_8\text{O} \rightarrow \,^{32}_{16}\text{S} + 16.5\ \text{MeV, etc.}$$

(c) After the oxygen has been used, the nucleus of the star again contracts and heats up. When the temperature reaches ~2 × 10^9 K, the photofission of nuclei will predominate. As a result of photofission free α-particles, p and n, again emerge and reactions that form heavier elements from silicon (the burning up of silicon) occur, i.e.,

$$^{28}_{14}Si + \gamma \rightarrow ^{24}_{12}Mg + ^{4}_{2}He$$

$$^{28}_{14}Si + ^{4}_{2}He \rightarrow ^{32}_{16}S + \gamma$$

Nuclei of the $A = 4n$ type will predominate since for nuclei heavier than silicon the gradual attachment of ^{2}He is preponderant. Thus ^{56}Ni and ^{56}Fe appear.

(d) Nuclei with A more than 100 cannot be formed during reactions with charged particles. It is assumed that they emerge in the gradual capture of neutrons by nuclei in the iron group elements. It is useful to distinguish between slow process (s) and rapid (r) reactions. The s process appears when the concentration of free neutrons is insignificant. The existence of s-process is proved by the clear correlation between the distribution of the elements (N) and the cross-section of radiational capture of neutrons $\sigma : \sigma \cdot N = \text{const}$.

(e) Nuclei following ^{83}Bi could not be formed by the s-process. There is some reason to suppose that most synthesis reactions for heavy elements take place within several minutes of massive star collapse (supernova outburst). Direct observations during supernova outburst confirm the synthesis of transuranium elements, i.e. the existence of r-process. The luminosity of the supernova No. GC 4725 had decreased within 600 days exponentially, with $T_{1/2} = 55$ days. This half-life is shared by 3 radionuclides: $^{7}_{4}Be$, $^{89}_{38}Si$ and $^{254}_{98}Cf$. If the luminosity were associated with the decay of ^{89}Si and $^{7}_{7}Be$, the abundance of their decay products $^{7}_{3}Li$ and $^{89}_{39}Y$ would be 100 times higher. This proves that the decay of ^{254}Cf did take place more rapidly.

REFERENCES

7.1. Baranov, V.I., Titayeva, N.A., 1973. Radiogeology. Moscow University Press, Moscow.

7.2. The Natural Isotopes of Hydrosphere, 1975. (Ed. V.I. Ferronskii). Nedra, Moscow.

7.3. Lal, D., 1972. Radionuclides—Cosmic Ray Produced.—Encyclopedia of Geochemistry and Environmental Sciences. (Ed. Fairbrüge). Van Nostrand, 996-1006.

7.4. Arslanov, Kh.A., 1987. Radiocarbon: Geochemistry and Geochronology. Leningrad University Press, Leningrad.

7.5. ^{10}Be and ^{9}Be in South Atlantic DSDP Site 515: Relation to Geomagnetic Reversals and to Sediment Composition, 1990.—Earth and Planet. Sci. Lett., **98**, No 314, 267-276.

7.6. Nishiizumi, Y., Klein, J., Middleton, R., and Craig, H., 1990. Cosmogenic ^{10}Be, ^{26}Al and ^{3}H in Olivine from Maui Lavas.—Earth and Planet. Sci. Lett., **98**, No 314, 263-266.

7.7. Kurz, M.D., Colodner, D., Trull, T.W., Moore, R.B., and O'Brien, K., 1990. Cosmic Ray Exposure Dating with *in situ* Produced Cosmogenic ^{3}H Result from Young Hawaiian Lava Flows.—Earth and Planet. Sci. Lett., **97**, No 1/2, 177-185.

7.8. Anderson, R.F., Lao, Y., Broecker, W.S., Trumbore, S.E., Hofmann, H.J. et al., 1990. Boundary Scavenging in the Pacific Ocean: A Comparison of ^{10}Be and ^{231}Pa.—Earth and Planet. Sci. Lett., **96**, No 3/4, 287-304.

7.9. The Isotopy of Waters in Nature, 1978. (Ed. V.I. Ferronskii). Nauka, Moscow, 244.

7.10. Managhan, M.S., Krishnawami, S., and Turecian, K.K., 1986. The Global-average Production Rate of ^{10}Be.—Earth and Planet. Sci. Lett., **76**, No 3/4, 275-287.

7.11. Amin, B.S., Lal, D., and Somayajulu, B.L.R., 1975. Chronology of Marine Sediments Using the ^{10}Be Method in Comparison with Other Methods.—Geochem. and Cosmochem. Acta, **39**, No 65, 73-103.

7.12. Nishiizumi, K., Lal, D. et al., 1986. Production of ^{10}Be and ^{26}Al by Cosmic Rays in Terrestrial Quartz *in situ* and Implication for Erosion Rates.—Nature, **319**, No 60, 134-139.

7.13. Tera, F., Brown L., Morris, J., and Sacs, S., 1986. Sediment Incorporation in Island-arc Magmas: Inferences from ^{10}Be.—Geochem. and Cosmochem. Acta, **50**, No 4, 535-550.

7.14. Lal, D., Nijampurkar, V.N., Somayajulu, B.L.K. et al., 1976. Silicon-32 Activities in Coastal Waters of the World Oceans.—Limnology and Oceanography, No 21, 285-293.

7.15. Amin, B.S., Kharkar, D.P. and Lal, D., 1966. Cosmogenic ^{10}Be and ^{26}Al in Marine Sediments.—Deep Sea Research, **13**, 805-824.

7.16. Magaritz, M., Kaufman, A., Levy, Y. et al., 1986. ^{36}Cl in a Halite Layer from the Bottom of Dead Sea.—Nature, **320**, No 6059, 256-257.

7.17. Shukolyukov, Yu.A., 1982. The Products of Heavy Nucleus Fission on Earth. Energoizdat, Moscow.

7.18. Lavrukhina, A.K., 1972. Nuclear Reactions in Cosmic Bodies, Nauka, Moscow.

7.19. Geochemistry and Cosmochemistry, 1984.—The 27th International Geological Congress. Rep., **11**, Nauka, Moscow.

8 Radionuclides of Anthropogenic Origin

As a result of the industrial development in the second half of this century, a great number of radioactive nuclides with anthropogenic origin have emerged. Their radioactivity on the average is commensurable with the natural radiation background, though in some territories it can exceed it significantly. Anthropogenic radioactive nuclides form two main groups. The first contains artificial radioactive nuclides that result from industrially induced nuclear reactions. They are not typical of the biosphere since they were practically absent on the Earth till the middle of the twentieth century.

The second group includes natural radioactive nuclides (uranium, thorium, and their decay products) whose concentrations on the Earth's surface has markedly increased due to industry.

8.1 The Environment and Its Pollution

The environment covers the habitat of human beings and animals, i.e. the biosphere. It is at the interface of the other terrestrial geospheres, such as the lithosphere, the hydrosphere, and the atmosphere. The biosphere is in close interaction with them. The interaction front is mainly the outer boundary of the lithosphere (the soil), which is in contact with the water and air layer of the Earth, provided the mutual penetration of geospheres. The exchange equilibrium between the biosphere and other geospheres helps to maintain its existence.

The technical progress of our civilization, and, particularly, its scientific and technical revolution in the second half of the twentieth century began to disturb the equilibrium between the geospheres. The use of natural resources and industrial production on a world-wide scale rapidly increased. According to estimates of the US Mining Bureau (1970), American industry alone used in 1940-1970 more raw materials than humanity had used up until World War I. This led to a vast amount of industrial waste, not ever handled by the biosphere, and to the contamination of the soil cover, the hydrosphere, and the atmosphere.

8.1.1 Natural Radiation Background and Anomalous Concentrations

Natural radioactive nuclides are part of the environment. The natural radiation background generated by them is the prerequisite for the existence of the biosphere, which emerged under the conditions produced by this background. Any substantial change in the radiation background will undoubtedly have a negative influence upon its inhabitants.

The gamma radiation on the Earth's surface produced by the natural background is mainly due to the rocks containing ^{232}Th with its decay products (~40%), containing ^{238}U and ^{235}U with their decay products (~25%) and ^{40}K(~35%). The concentrations of uranium, thorium, and potassium can be

Table 8.1 Average Individual Radiation Doses for the Population in the USSR in 1980-81 (μSv/year) [8.3]

Sources of irradiation	Type of irradiation	Effective equivalent dose
Natural sources	External	620
	Internal	350
	Total	1000
Technogenic radiation background:	External	1
building materials (additional irradiation in the	Internal	950
buildings)	Total	1050
	External	3.5×10^{-2}
Fertilizers	Internal	10.1×10^{-2}
	Total	13.6×10^{-2}
Coal-fired power	External	5×10^{-2}
stations with	Internal	1.9
capacity 7 GW	Total	2.0
X-ray examinations	External	1400
Global radioactive	External	10
fall-out after	Internal	13
nuclear explosions	Total	23
Nuclear power sta-	External	9×10^{-2}
tions with capacity	Internal	8×10^{-2}
over 12 GW	Total	17×10^{-2}
Others	Total	1.6
Total impact		3500

assessed from the average content of these elements in the soils of the USSR (according to Boltneva et al., 1980), i.e.,

$$U = 1.5 \text{ ppm}; \quad Th = 5.5 \text{ ppm}; \quad K = 1.2\%$$

The average population dose rate from external source of the natural radionuclides is 4.6 μrad/h, while from cosmogenic radionuclides it is 0.02 μrad/h, and from the ionizing component of cosmic radiation at sea level it is 3.15 μrad/h [8.1, 8.2].

Variations in the concentrations of radionuclides on the surface of the lithosphere lead to substantial changes in the natural radiation background. Classical examples of regions with higher natural radiation backgrounds are the States of Kerala and Tamil Nadu in India and the Atlantic coast of Brazil. This is where monazitic littoral-sea placers lie over enormous areas (Ch. 3).

In India one part of the ocean coast covers 55 km and has a population of 70 000 people, the average absorbed dose reaches 130 μrad/h, while in the State of Kerala it is 380 μrad/h. In Brazil, in the States of Espírito Santo and Rio de Janeiro, the absorbed dose in the city streets can be as high as 100-200 μrad/h, while on the beaches it is 2000 μrad/h [8.1, 8.2].

The radiation background from industrial sources on a global scale arises due to various factors which will be dealt with in more detail later. The contribution of different sources of radiation to the average dose received now by a person is illustrated in Table 8.1.

8.1.2 Effects of Ionizing Radiation on Living Organisms [8.1-8.6]

The biological effect of ionizing radiation comes from the ionization and excitation of atoms and depends on how great the radiation dose is (see Ch. 1). Depending on the dose and exposure time acute radiation injury due to large single doses (more than 0.5 Sv) is distinguished from chronic injury due to repeated radiation with small dose (less than 0.5 Sv). Acute exposure is responsible for radiation sickness and, in the end, can be lethal. Chronic exposure with large doses (more than 10 Sv) can cause a chronic radiation sickness and damage to individual organs.

Affliction manifested after a long time after a single dose or chronic irradiation is called aftereffect. The effect is determined by the total accumulated dose irrespective of whether it was received in one day or 50 years. An assumption is made on the thresholdless linear dependence of the dose-effect relationship. Individual afflictions are the consequences of small doses and are called stochastic. The two principal groups of stochastic afflictions are (a) somatic-malignant tumours, and (b) genetic malformations and defects.

At present we believe continuous radiation in doses of up to 0.05 Sv a year do not bring about any somatic changes in the organism of a healthy adult. The harm caused by genetic consequences is approximately 1/3 of the cancerogenic effect. To prevent the biological effects of radiation it is necessary to fulfill fairly rigid norms and rules.

A special topic is the effect of substantial doses of ionizing radiation on a biogeocenosis in general (in cases of fallout during nuclear explosions, big accidents at the nuclear power stations, etc.).

Biogeocenosis is defined as the initial unit of the community of living organisms in the biosphere that dwell in a limited domain on the earth's surface. A biogeocenosis has homogeneous natural conditions (rocks, soils, climate, geomorphological and hydrochemical conditions, etc.) and is a multicomponent system, the normal functioning of which is maintained by properly balanced direct and indirect bonds between its inhabitants (animals, plants, and microorganisms).

The effect of ionizing radiation on a biogeocenosis can be seen by the emergence of primary changes connected with a different sensitivity of organisms to radiation, and the secondary variations are brought about by primary (initial) modifications [8.1, 8.2].

For example, the lethal dose in case of acute irradiation for the most livestock fluctuates from 3 to 15 Gy, and 25-30 Gy for poultry. Maximum radioresistance (stability) is shared by insects. The fruit fly can stand a dose of 850 Gy. Among the vegetative organisms the least resistant to radioactivity are wood plants, particularly conifers [50% of the lethal dose (LD) = 12 Gy].

At the same time the lower plants (ferns, mosses, lichens, and mushrooms) can stand doses up to 10 000 Gy. The primary radiation effect depending on the radiation dose, can cause the retardation of growth and development, up to the organism death.

Secondary effects are due to a disturbance of the bonds between the members of biogeocenosis as a result of the depression or death of the least radio-resistant organisms. This results in a release of competitive stress from more stable species that have been earlier depressed which begin to develop intensely.

In some species the radiation can stimulate biochemical processes, enhancing their competition in biogeocenosis. Thus, the disturbance of biological equilibrium due to irradiation in nonlethal doses can lead to a complete change in the biogeocenosis.

8.2 The Pollution of the Environment by Artificial Radionuclides

Artificial radioactive nuclides appeared on Earth in the middle of twentieth century as a result of nuclear reactions carried out by man. These radionuclides obviously existed in the protoplanetary cloud, but they readily decayed due to their short half-lives.

Nowadays there are two main sources of environmental contamination by artificial radioactive nuclides, viz., nuclear arms (nuclear tests) and nuclear reactors.

8.2.1 Pollution of the Environment from Nuclear Explosions

The principal sources of environmental radioactive pollution are nuclear explosions, which have been carried out since 1945.

Nuclear explosions are classified by the type of reaction (fission, fusion), power and location of explosion (stratospheric, tropospheric, underground, and underwater). Each explosion is accompanied by the formation of an enormous number of radioactive nuclides: fission products, activated container materials and the environment, and the remains of unreacted nuclear fuel [8.3].

A nuclear explosion with the power equivalent to 1 kton of TNT corresponds to 1.45×10^{23} fission events. We have therefore the formation of about 70 fragmental radioactive nuclides, which initiate chains of radioactive transformations consisting of 2-6 units. An hour after explosion there will be more than 130 radioactive nuclides, of which about 30% have half-lives of one hour, 26% of one day, 32% from a day to a year, and 5% more than one year [8.6]. The total activity of the fragments is proportional to the power of the explosion, and, as is the case with the composition of fragments, slightly depends on the type of the material undergoing fission (^{235}U, ^{239}Pu) (Table 8.2). The composition of unreacted fuel is determined by the construction of the installation. For instance, if the fuel is ^{239}Pu, then during the bomb explosion of 20 kton the surrounding medium will receive about 3 Ci of ^{239}Pu [8.7].

The induced radioactivity of the installation material and of the environment is markedly lower than that of fission fragments. As a result of the irradiation of construction material we get ^{22}Na, ^{54}Mn, ^{57}Co, ^{58}Co, ^{60}Co, ^{59}Fe, ^{65}Zn, ^{181}W, ^{185}W, ^{187}W, and ^{203}Pb. When neutrons react with atmospheric nitrogen, ^{14}C is formed, while with the soil, we get ^{28}Al, ^{24}Na, ^{56}Mo, ^{59}Fe, and others. During explosions in the ocean the activation products will correspond to the composition of sea water (Table 8.2). The concentration of an activation product can be calculated by [8.7]:

$$N_i = \Phi \frac{n_i \sigma_{i,a}}{\sum n_i, \sigma_{i,n}}$$

where Φ is the total number of neutrons formed during the explosion, n_i is the concentration of the element in the irradiated volume of water; $\sigma_{i,a}$ is the cross-section of the activation capture of neutrons, $\sigma_{i,n}$ is the cross-section of the absorption of neutrons which does not lead to the formation of radionuclides. The degree of contamination of the Earth's surface by the radioactive products of nuclear explosion depends on its character. The contamination of vast areas takes place during atmospheric nuclear explosions, as well as when not very deep underground or underwater explosions are performed, which are accompanied by the ejection of ground or water. In cases of underground nuclear explosions without the ejection of ground most of the radioactive nuclides stay inside. Only isotopes of crypton and xenon can diffuse along the fissures and get into the atmosphere. The pollution of underground waters cannot be excluded.

Table 8.2 Initial Activities of the Main Fission Products (Ci per 1 kton of Power [8.8])

Radionuclide	^{235}U	^{239}Pu
$^{90}Sr + ^{90}Y$	1.2×10^2	6.0×10^1
$^{95}Zr + ^{95}Nb$	3.3×10^4	2.6×10^4
^{103}Ru	3.2×10^4	4.9×10^4
$^{106}Ru + ^{106}Rh$	4.0×10^2	5.5×10^3
^{125}Sb	2.6×10^1	5.5×10^1
^{131}I	1.2×10^5	1.9×10^5
$^{132}Te + ^{132}I$	4.4×10^5	6.5×10^5
$^{137}Cs + ^{137}Ba$	1.8×10^2	1.8×10^2
$^{140}Ba + ^{140}La$	1.4×10^5	1.2×10^5
^{141}Ce	5.0×10^4	4.2×10^4
$^{144}Ce + ^{144}Pr$	5.5×10^3	4.1×10^3

Atmospheric explosions are accompanied by the emergence of the evaporated substance in the form of a great sphere, which elevates up as a result of expansion and buoyancy.

When the sphere cools up the radioactive gases begin to condense with the formation of particles which are then coagulated. The first to condensate, after several seconds of the explosion are the oxides of the most refractory metals (Al, Fe, Mg, and others), which form particles from 0.4 μm to 4.0 μm in diameter. The separation of refractory compounds is followed by the condensation of less refractory components containing relatively volatile radioactive nuclides and those which have volatile predecessors. This part of the radioactive cloud yield finer aerosols (< 0.4 μm). The gravitational sedimentation of submicrone aerosol particles is insignificant. They move with the air masses and adsorb on natural atmospheric aerosols. In this case no local contamination occurs.

During explosions near the earth's surface, the radioactive cloud contains, in addition to the evaporated particles from the ground, a large number of melted mineral particles ranging in size from a micron to several millimeters, with radionuclides on their surface.

Up to 80% of the radioactivity is carried away by large particles. This results in the territories with heavy local pollution and a high level of radiation. All the radionuclides that form during a nuclear explosion are referred to as the radioactive fallout. Local, tropospheric, and stratospheric fallouts are distinguished. The local fallouts contain particles with diameters of more than 10 μm. They stay in the atmosphere for less than a few days, usually for some hours. The local fallout is restricted by a relatively narrow trace of radioactive cloud. The trace may extend to hundreds or sometimes thousands of kilometres.

Tropospheric fallout consists of small aerosols (from 1.0 to 10 μm), which could not penetrate through the tropopause. Their average half-life period is about 30 days. Within this period of time they manage to travel around the Earth several times. Tropospheric fallouts mainly contain radionuclides with half-lives of several days (e.g. ^{131}I, ^{140}Ba, ^{89}Sr). The 'blurring' of fallouts along the latitude ranges within 15°-20°.

Stratospheric fallouts are formed by particles that are at least 1 μm in size. After being moved into the stratosphere, radionuclides rapidly disperse throughout the hemisphere. The half-life period of most of them from the stratosphere may take several years. Gaseous radionuclides are retained in the atmosphere even longer. Thus, the stratosphere is a kind of special storehouse for radionuclides with micron and submicron dimensions.

The transfer of radioactive nuclides from the stratosphere into troposphere most likely takes place via the subtropical breakout of the tropopause at the expense of horizontal movement and reaches its maximum at the end of winter. Spring global fallouts of radionuclides are found to be greater than the average annual level. Stratospheric fallouts are the source of contamination by the more long-lived radionuclides (e.g. ^{90}Sr, ^{137}Cs, ^{144}Ce, ^{14}C, and ^{3}H). Stratospheric fallout is often referred to as global fallout. Radioactive fallouts come out on to the earth's surface from the troposphere as a result of three processes: (1) the formation of droplets in the clouds; (2) due to washing out by the falling raindrops; and (3) at the expense of dry sedimentation, which is particularly important for arid regions.

8.2.2 Nuclear Reactors and Environmental Pollution

The role of nuclear energy for power stations in the world's overall energy balance is constantly increasing and, according to some estimates will provide nearly 50% of the general power consumption by year 2020 [8.3]. This is because the reserves of organic fuel, fossil fuels have begun to be dismantled and the main energy supply rests on the nuclear power stations. These stations are a part of a fuel-energy complex amalgamating all the enterprises of the nuclear fuel cycle, viz. mining and processing radioactive ore; the production of fuel elements; the nuclear reactor stations which use the fuel; the reprocessing plants for spent fuel elements; storage facilities for radioactive wastes; and the transportation of radioactive material between stages. The artificial radionuclides emerge here only from the reactor.

A nuclear reactor is a device in which the controlled chain nuclear reaction takes place. Contemporary industrial reactors only operate using the fission reaction. There are two basic types of reactors: those that function with thermal neutrons and those that work with fast neutrons.

The thermal neutron reactors are the most widely used. The ability of fissioning after being hit by a thermal neutron is shared by ^{235}U, ^{233}U, ^{239}Pu, and ^{231}Pu. Only one fissioning radionuclide ^{235}U is found in nature and it makes up 0.71% of natural uranium. For a reactor to operate it is necessary to enrich the uranium with the isotope ^{235}U, as nuclear fuel requires 1.8%-4.4%. The enriched uranium, in the form of UO_2 tablets, is placed into hollow hermetic rods, or the fuel elements, made of zirconium alloys. Since the cross section of the ^{235}U fission reaction is maximum for thermal neutrons, the fuel elements are put into a neutron moderator, usually heavy water or graphite. The principal part of the reactor is the active zone, where the fuel elements are situated and the nuclear reaction takes place. The heat generated by the exothermic fission reaction is carried away from the active zone by a liquid or gaseous coolant (e.g. H_2O, CO_2, Na, etc.), which may also be the moderator (water). The nuclear reaction is controlled using rods containing materials that absorb neutrons (e.g. boron or cadmium). The rods can be inserted into the active zone and either retard or terminate the reaction. The active zone together with a neutron reflector is enclosed in a protective casing.

Nuclear reactions using fast neutrons do not contain any moderator. The reflector is made of materials that are not fissionable by thermal neutrons, though it can breed nuclear fuel (^{238}U and ^{232}Th). The reactions in the reproduction zone involve fast neutrons: $^{238}U + n \rightarrow {}^{239}U \rightarrow {}^{239}Np \rightarrow {}^{239}Pu$; and $^{232}Th + n \rightarrow {}^{233}Th \rightarrow {}^{233}U$. They give rise to the fissionable radionuclides ^{239}Pu and ^{233}U. As distinct from thermal neutron reactors, where 1-2% of nuclear fuel is utilized, fast neutron reactors 20-30 times more energy is produced.

Nuclear reactors produce solid, liquid and gaseous radioactive wastes, namely, the fission products from the nuclear fuel, activization products, and unreacted fuel [8.3].

The fission products mainly form within the fuel elements and remain there. Some leave these elements, but only in infinitesimal amounts, except tritium (^{3}H). About 80% of ^{3}H penetrates stainless steel shells, while 99% of the escaping ^{3}H binds to the zirconium. Small cracks and defects in the shells of fuel elements allow the coolant to be contaminated. The fission products include noble gases, volatile and nonvolatile substances (Tables 8.3 and 8.4). All are beta- and gamma emitters and have half-lives ranging from fractions of a second to decades.

Activation products include radionuclides with various half-life (Tables 8.3 and 8.4). The most long-lived are the products of fuel activation (isotopes of the transuranic elements Np, Pu, Am, and Cm), which have half-lives of some tens of thousands of years. The less long-lived come from activation of the coolant, and construction materials.

Table 8.3 Biologically Significant Radionuclides Formed in a Nuclear Reactor [8.5]

Nuclide	$T_{1/2}$	Nuclide	$T_{1/2}$	Nuclide	$T_{1/2}$
		Noble gases and iodine			
^{85}Kr	10.7 years	^{133}Xe	5.2 days	^{129}I	1.6×10^{7}years
^{85m}Kr	4.5 hours	^{133m}Xe	2.2 days	^{131}I	8 days
^{87}Kr	1.3 hours	^{135}Xe	9.1 hours	^{133}I	21 hours
^{88}Kr	2.8 hours	^{135m}Xe	15.7 min	^{135}I	6.6 hours
		Solid products of fission			
^{89}Sr	51 days	^{103}Ru	39 days	^{140}Ba	13 days
^{90}Sr	28.6 years	^{106}Ru	1 year	^{141}Ce	33 days
^{91}Y	59 days	^{129}Te	34 days	^{143}Pm	14 days
^{95}Nb	35 days	^{134}Cs	2.1 years	^{144}Ce	284 days
^{95}Zr	64 days	^{137}Cs	30 years	^{155}Eu	5 years
		Activization products			
^{51}Cp	28 days	^{59}Fe	45 days	^{3}H	12.3 years
^{54}Mn	312 days	^{65}Zn	244 days	^{14}C	5730 years
^{58}Co	71 days	^{95}Nb	35 days	^{41}Ar	1.8 hours
^{60}Co	5.3 years	^{95}Zr	64 days	^{110}Ag	250 days

Table 8.4 Activity of Fission Products and Transuranic Elements in Spent Fuel

Radionuclide	Half-Life	Specific activity in Ci/ton of fuel	
		thermal reactor	fast neutron reactor
		Fission products	
^{144}Ce	284 days	892	1280
^{95}Nb	35 days	869	2660
^{95}Zr	64 days	524	2100
^{106}Ru	1 year	459	1290
^{89}Sr	51 days	214	637
^{137}Cs	30 years	107	109
^{147}Pm	2.6 years	104	353
^{90}Sr	29 years	76.9	43.4
^{3}H	12.3 years	0.7	0.9
		Transuranic elements	
^{241}Pu	15 years	116	600
^{242}Cm	163 days	19.3	65.5
^{238}Pu	88 years	2.79	11.2
^{244}Cm	18 years	2.50	1.24
^{240}Pu	6.5×10^3years	0.48	0.43
^{239}Pu	2.4×10^4 years	0.33	3.53
^{241}Am	433 years	0.17	1.57
^{243}Am	7330 years	17×10^{-3}	50×10^{-3}
^{242}Pu	3.8×10^5years	1.4×10^{-3}	13×10^{-3}

Most of the radioactivity remains within the fuel elements. Spent elements are kept within the nuclear power station in special pools where the short-lived radionuclides decay.

The fuel rods are then delivered to a reprocessing plant where they become the waste of the final stage in the nuclear fuel cycle.

A nuclear power station is designed so that the radionuclides entering the environment do not exceed the permissible concentrations set by health authorities.

Gaseous wastes are mainly formed when cleansing the coolant, due to leaks in pipelines and ventilating air. Gaseous wastes undergo a complex system of purification in filters and chromatographic columns. Sometimes gases are held in reservoirs for several days or weeks and their activity decreases. After the cleansing process the gases are ejected through a tall chimney (100-150m) to ensure good dilution in the atmosphere. The ejected wastes include the radioactive noble gases (the isotopes of krypton and xenon), small quantities of tritium, and occasionally some aerosols. The ejections of ^{3}H are of the order of 20 Ci/MW of electric power a year; while with ^{14}C they are 6-10 mCi/MW a year [8.3].

Liquid wastes are not discharged directly into the environment. They are subjected to an elaborate purification in chromatographic columns and sedimentors, etc. The purified water is used again, while the solid concentrate is placed in tanks and buried after solidification. When the purified water tanks are full they can be partially vented into the environment. This is debalanced water, whose activity does not exceed the maximum permissible concentration for drinking water. In the main the activity comes from tritium which is not removed by the water purification. Solid wastes with low specific activity occupy relatively large volumes, while those with high activity that were accumulated in purification of liquid and gaseous wastes make up only several percent. Lower activity waste is buried on-site in power station, while very active waste is stored in special facilities.

Reprocessing Irradiated Fuel

Spent fuel elements are delivered to a reprocessing plant to extract useful components and transform waste into a convenient form for burying. According to [8.3] reprocessed fuel mainly contains fission products and transuranium elements. Thus, it contains over 99% of ^{85}Kr, ^{3}H, and ^{129}I, as well as approximately 2/3 ^{14}C, the activation product. The environmental pollution due to reprocessing plants is much more serious than that by nuclear power stations.

Today the plants for reprocessing nuclear fuel do not have large throughputs. Reprocessing plants eject much more gases than typical nuclear power stations, and so their exhaust chimneys are as high as 200 metres.

On the whole the most tangible contribution to global pollution by the nuclear fuel cycle comes from ^{14}C, ^{85}Kr, ^{3}H, and ^{129}I. Other long-lived non-gaseous radionuclides (^{90}Sr, ^{137}Cs, and the isotopes of the transuranium elements) are ejected into the environment by these enterprises and scattered over vast territories.

Accidents at Nuclear Power Stations

In addition to 'planned' ejections, another source of contamination at nuclear power stations are accidents. The main hazard is from fission and activation products accumulated in the active zone. They are preserved within the fuel elements as long as the elements and their casing remain sealed and their cooling is sufficient. The fuel can be overheated and melt only when the heat flux evolved during the nuclear reaction exceeds the value removed by the coolant. This can occur if there is disturbance in the primary circuit and the coolant is lost. In addition the active zone may overheat and melt. In this zone the coolant is vaporized leading to excess pressure and explosion. If the protective can is destroyed, the fission and activation products together with fuel that has not undergone fission will enter the environment and pollute it, the contaminated area depends on the explosion and meteorological conditions. The radionuclides will include the noble gases (Kr, Xe); halogens (I, Br), alkali metals (Cs, Rb); rare-earth elements (La, Ce, Pr, Nd, Pm, Sm, Eu), Y, as well as Ru, Rh, Pd, W, Mo, Ag, Tc, Se, Te, S, Sr, Ba, Cr, Np, Cm, Pu.

Accident at Chernobyl Atomic Power Station [8.9-8.12]

The greatest accident of the atomic reactor occurred at Chernobyl atomic power station in April 26, 1986. The accident has caused the partial damage to the active zone of reactor and the system of cooling in the fourth

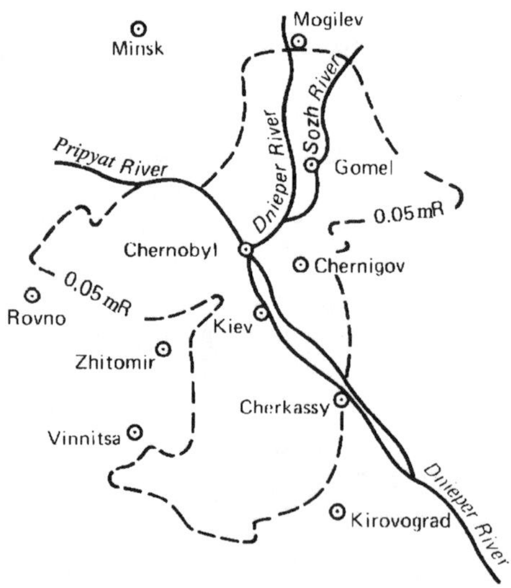

Fig. 8.1. Conditional pollution boundaries as a result of the Chernobyl atomic power station disaster. The dotted line corresponds to the level of radiation of 0.05 mR/h for June 10, 1986 [8.10]

block. Under these conditions, an increased heat evolution in a reactor shaft has occurred due to residual heat release of fuel during the decay of fission products and to various chemical reactions (burning of H_2, oxidation of graphite and ziconium, etc.). At the moment of explosion the fuel was heated up to 1600-1800 K and dispersed fuel, graphite and the products of its oxidation, enriched with the products of fission discharged into the atmosphere. First, the volatile products (isotopes of noble gases, I, Cs, Te) were predominantly evolved and then the composition of ejections was close to radionuclide composition of fuel fragments. As a result, considerable contamination of the vast territory occurred.

Figure 8.1 shows the contamination boundaries by 10 June, 1986, corresponding to a conventional level of 0.05 mR/h which exceeded the natural background 5-10 times. Before the accident, the natural background in the region of Chernobyl atomic power station was 0.01-0.015 mR/h and the density of global contamination was 0.10 Ci/km^2 (^{137}Cs) and 0.07 Ci/km^2 (^{90}Sr). Data of aeroradiometric exploration show that the discharge in the boundaries of contaminated zone included in (mCi): ^{239}Np–720, ^{99}Mo–160, ^{132}Te–73, ^{131}I–86, ^{140}Ba–135, ^{141}Ce–150, ^{103}Ru–130, ^{89}Sr–63, ^{91}Y–75, ^{95}Zr–130, ^{95}Nb–130, ^{144}Ce–90, ^{106}Ru–60, ^{125}Sb–4.0, ^{90}Sr–6.0. The contaminated territory has an irregular form and inhomogeneous distribution of activity and radionuclide composition. For instance, the northern part is enriched with ^{137}Cs(+^{134}Cs). This is due to dynamics of discharge and meteorological situation, rains, in particular, at the moment when radioactive cloud is passing. There were places where Cs content reached 20-30 Ci/km and even 80 Ci/km, while Pu content was 0.1-1.0 Ci/km.

Water ecosystems were contaminated by aerosols formed from the jet of the discharge on the first days of accident and then owing to washout by rains from contaminated water catchment area. In the period of fallouts, and after them, the concentration of radionuclides increased three to four times. The waters of the Pripyat River were mostly contaminated: when the aerosol fallouts stopped to form the overall beta activity of water decreased from 10^{-7} Ci/l to 10^{-9}-10^{-10} Ci/l (July-August, 1986). Concentration of ^{137}Cs in Pripyat at the end of May, 1986 was $1.1 \cdot 10^{-9}$ Ci/l (global background being $2 \cdot 10^{-10}$ Ci/l). The ^{90}Sr content was $4 \cdot 10^{-10}$ Ci/l during the first days after the accident and today it is about $(3\text{-}5) \cdot 10^{-11}$ Ci/l. About 70 per cent of ^{90}Sr precipitates on a river suspended matter. The ^{137}Cs/^{90}Sr ratio before the accident was 0.1, and after the accident it increased 100 times.

Within the first two or three months after the accident, the contamination of the territorial ecosystems involved the upper 1-2 cm layer of a litter, shed and moss. Further, migration into the soil has been noted. The influence of contamination on plants has been manifested in morphological changes, most severe when the doses are 100-300 rad. The pine forests were more susceptible to radiation. The territory of damaged forest was 400 ha (D > 1000 rad). In agricultural plants, roots were not firstly contaminated (as a result of aerosol fallouts), and then the whole plant was contaminated (as a result of wind).The maximum coefficient of radioactive transition from the ground was observed for grass and winter rye. The contamination of pastures has resulted in ^{131}I, and ^{137}Cs + ^{134}Cs penetration into milk and meat.

The Burial of Radioactive Wastes

The development of the nuclear industry leads to an ever-growing amount of radioactive waste. Its removal from the biosphere is a complicated technological and geochemical task.

Wastes can be subdivided into three categories according to the level of radioactivity, viz.:

(1) wastes with low specific activity ($< 1 \cdot 10^{-5}$ Ci/l);
(2) wastes with average specific activity (from $1 \cdot 10^{-5}$ to 1.0 Ci/l);
(3) wastes with high specific activity (> 1.0 Ci/l).

Group 1 waste is hazardous only when it penetrates an organism; group 2 waste is a source of external irradiation; and group 3 waste also generates heat. Wastes of groups 1 and 2 are formed in the first stage of nuclear fuel cycle and will be treated below. The radioactive products formed in a reactor belong to group three. They are mainly liquid wastes of reprocessing fuel. There are several methods for extracting U and Pu from irradiated fuel. Usually countercurrent extraction is used for uranium and Pu by tributyl phosphate from the nitrate solution. The residual nitric solution containing 99.9% of the fission products, some of the nonextracted Pu, and the transuranium elements (mainly Np, Am, Cm) (Table 8.4) is buried.

In the developed countries the following procedure has been adopted to deal with group 3 waste:

(1) storage in liquid form until the residual heat evolution is lowered;
(2) solidification and storage in controlled conditions;
(3) final burial of solid waste in stable geological formations.

The first operation lasts five years, while the second has to be accomplished before ten years.

Solidification involves calcination with the consequent melting, or vitrification of the waste. A modern method is to melt the wastes into glass blocks which are then placed in metal capsule and buried. Experiments have revealed that the glass can be affected by water with 2×10^{-5} to 2×10^{-7} g/(cm^2 $\times$ days) being leached. On the average this corresponds to the removal of 1.5 mm of glass in 100 years. Solidification of group 3 waste is being done by forming ceramic, metalloceramic and glass-metal blocks, which have a high thermal conductivity. At the same time their mechanical, chemical, and radiation stability is considered.

Nowadays radioactive wastes is buried in deep geological formations. Other projects exist e.g. burial in deep trenches on the sea floor, below the seabed, the ejection of waste by rocket beyond earth orbit, irradiating long-lived radionuclides in a reactor to transmute them. These are either less reliable or more expensive than conventional burial. When underground storage facilities are constructed in mines, natural barriers should be supplemented by artificial ones to diminish the possibility of radionuclides migrating into the biosphere. The most important conditions are [8.3]:

(1) the physical chemistry of the waste should make it resistant to migration.

(2) corrosive stability and a prolonged service period of the capsule material into which the waste is placed.

(3) filling of chambers and tunnels with buffer materials that have a low permeability and high sorption.

(4) long period of time within which the ground water comes to the nearest consumer.

(5) the use of rock with a low permeability and low rates of underground water, or permeable self-healing rocks.

(6) the sorption delay of radionuclides if there is no migration along cracks in the rock or in overlying rocks; the absence of useful mineral deposits nearby.

(7) the depth of storage depending on the lithostatic loading should provide the integrity of the storage and the heat conductivity of rocks must ensure heat removal.

1. Burial in rocks is popular in many countries. The advantage is that the rock is stable and waste can be withdrawn when necessary, while the drawback is the inability to selfhealing of cracks as happens in saline and clay formations. The best conditions are storage in granites and gneisses. Special attention should be paid to the hydrogeological conditions. The depth of burial should be of the order of 100-500 metres. In Sweden a burial is planned in the rocks at about 500 m in granite-gneisses, where the age of water is about 10 thousand years, while its velocity is 0.1-0.2 $l/m^2 \cdot year$. The buffer will be bentonite, which will increase the pH of the fissure waters to 8-9 and will reduce the rate of underground water movement by 200 times. To avert corrosion it is planned to put the steel cylinders with borosilicate glass and group 3 waste into sealed lead containers with walls 100 mm thick and an external titanium covering 6 mm thick to prevent the radiolysis of the underground waters.

2. The burial of group 3 waste in saline formations has several advantages, i.e. adequate heat conductivity, aqueous impermeability, the absence of water, plasticity, and the lack of cracks. When designing this kind of storage it is important to estimate the radiogenic thermal field which could affect the safety of the storage facility.

3. Burial in clay rocks on land and under the seabed is the policy in a number of countries. Other methods include sinking of containers in deep depressions directly on the ocean floor. Experiments and estimates indicate that the velocity of leaching radionuclides by sea water from vitreous blocks is of the order of 10^{-5} g/cm^2·days.

The period over which waste with a high specific activity can be preserved is determined by the decay period of the most long-lived ones. If the wastes do not contain any transuranium elements, their activity after 500-600 years will not markedly exceed the background level. The presence of transuranics requires longer and more reliable preservation lasting tens of thousands of years. The separation of trivalent Am and Cm from the rare-earth elements is still a fairly complicated technology.

8.3 The Migration of Artificial Radionuclides

8.3.1 Migration in Sea Water

Radionuclides get into the ocean from the atmosphere with continental run-off (including the direct discharge of waste) and due to reactors aboard atomic vessels. For example, tritium, which exists in the atmosphere as water vapour in the form of $^1H^3HO$ readily undergoes isotopic exchange with the surface layer of the ocean. The concentration of tritium in this layer is given in Fig. 8.2. ^{14}C, which is present in the troposphere in the form of $^{14}CO_2$, also enters the ocean as a result of isotopic exchange. The distribution of anthropogenic ^{14}C in the surface layer of the ocean is shown in Fig. 8.3. Isotopes of Sr, Cs, Ce, Ru, Zr, Nb and of other non-volatile elements enter the atmosphere in the form of aerosols, Sr isotopes having the greatest solubility.

A major factor governing the pollution of the ocean is the discharge of radioactive waste. The chemical composition of this kind of waste and the forms in which the radionuclides exist greatly depend on the technology at the enterprise. They may be acidic or alkali solutions, concentrated salt solutions, or contain the organic compounds used in the extraction processes. Radionuclides can be found in the form of complex compounds with organic and inorganic ligands [8.7]. Zirconium, for instance, can exist as a stable complex with fluoride and oxalate ions. At the same time when there is low acidity, small concentrations of Zr^{4+} will give rise to colloids. The Nb^{5+} formed during the processing of nuclear fuel also readily hydrolizes with the formation

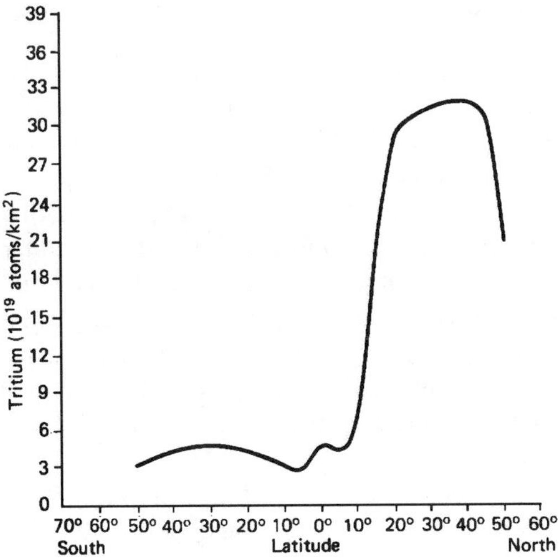

Fig. 8.2. Latitudinal distribution of ^{3}H per unit area in the upper layer of the Pacific Ocean to the depth of 500 m in 1965-1972 [8.2].

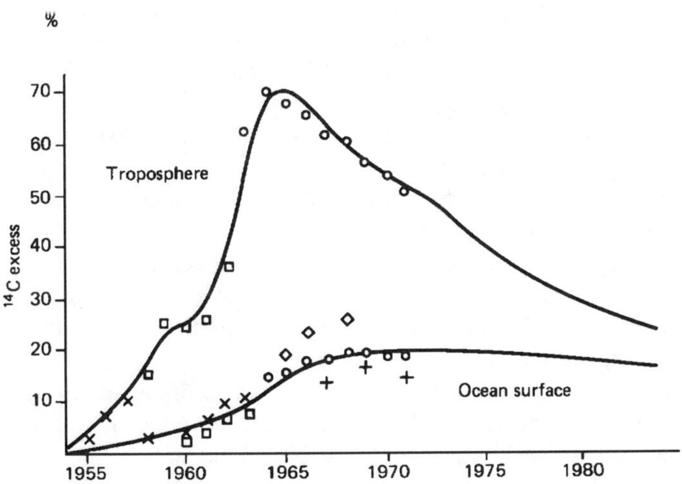

Fig. 8.3. Fluctuations of ^{14}C excess in the troposphere and on the ocean surface in 1955-1980 [8.2].

of colloids. Technetium, which has several oxidation states, easily oxidizes to +7, thus forming the anion TcO_4^- typical in aqueous media. In sea water ruthenium, as a rule, forms stable complexes. In acid solution it is found in the form of the cations Ru^{3+} and Ru^{4+}. In alkali media ruthenium is more usually in an anionic form, where its valency is + 5, +6, +7. In liquid discharges, strontium and caesium are usually in the form of simple ions Sr^{2+} and Cs^{1+}. Iodine actively reacts with organic extractants to form stable compounds. The rare-earth elements are usually present as hydroxides (in alkali media) or cations (in acid media). Uranium, plutonium and thorium also usually exist as simple or complex ions. Thus radionuclides enter the ocean with liquid waste mainly in the form of soluble components and become immediately involved in all the physicochemical and biochemical processes that take place in aqueous medium.

Hence, the same radionuclides can be found in sea water in the form of components that are different in composition and solubility. They will be markedly different from the natural elements (isotopic carriers) dissolved in sea water.

Three main groups of radionuclides are distinguished according to their behaviour in sea water:

(1) those that are easily soluble in water (^{3}H, ^{99}Tc, ^{90}Sr, ^{89}Sr);

(2) radionuclides that are sorbed on suspension (rare-earth elements, ^{91}Y, ^{95}Zr, ^{95}Nb, ^{106}Ru, and partially Pu, and some other induced varieties);

(3) biologically active radionuclides (induced radionuclides from the iron group (viz. ^{54}Mn, ^{60}Co, ^{55}Fe, ^{63}Ni, ^{51}Cr, ^{65}Zn, and plutonium). Group (1) elements migrate with the circulation of the water mass, group (2) elements are affected by sedimentation processes, and group (3) elements migrate depending on the behaviour of the biomass.

The geochemistries of ^{90}Sr and ^{137}Cs are determined by the behaviour of their isotopic and non-isotopic carriers.

Thus, most ^{90}Sr is contained in the perfectly mixed layer of sea water above the thermocline and only about 1% may have now passed into bottom sediments. In littoral regions and estuaries the ^{90}Sr concentration is at least five times greater due to the runoff from land.

The accumulation coefficient of ^{90}Sr by marine organisms is 100 for algae, 10 for crabs, and 1 for fish. Most ^{137}Cs is concentrated in surface water, and is 30 times smaller in deep layers. The biological half-life for ^{137}Cs from the surface layer lasts nearly 17 years. The accumulation coefficient by sea organisms is about 50 years.

Plutonium isotopes do not have isotopic or non-isotopic carriers in sea water, as ^{137}Cs and ^{90}Sr do and so they sink to a greater depth. More than 50% of plutonium is found below the thermocline and up to 36% is concentrated in bottom sediments. The absorption coefficient of Pu from water by algae, plankton and crustaceans ranges from 1000 to 3000, while for fish it is between 1 and 10.

8.3.2 Migration on Land

Most artificial radionuclides enter the land from the atmosphere as global fallout and as the discharge of enterprises in the nuclear cycle or as waste from them.

Radionuclides from the atmosphere rapidly begin to react with the upper layers of soils and migrate due to the processes in the soil and vegetation cover. Soil is a complex polydispersed system, within which there are several fairly homogeneous subsystems: (1) coarse grain particles, or the fragments of the original minerals of rocks; (2) fine grain particles, i.e. secondary clay minerals, humus, etc.; (3) gel films covering the particles and consisting of Fe, Mn, Al oxides, silicic acid, organic compounds, salts, etc.; (4) plant roots and organisms; (5) soil solutions; (6) and soil gases.

The behaviour of the radionuclides that come to the soil surface will depend on their chemistry, forms of existence, concentrations, the forms in which isotopic and non-isotopic carriers occur, climate and the composition of the soil, the chemistry and physical chemistry of the medium, and features of the landscape.

The absorption of radionuclide particles, or their fission products, by the same soil grows in the series: $^{106}Ru < ^{90}Sr < ^{144}Ce < ^{137}Cs$, and their bond stability grows as $^{90}Sr < ^{106}Ru < ^{95}Zr < ^{144}Ce < ^{137}Cs$ (Pavlotskaya, 1974). This is largely due to differences in the factors governing absorption. For example, the ion-exchange mechanism is predominant for strontium ^{89}Sr and ^{90}Sr isotopes, which are exchanged with stable Sr and Ca isotopes in the soil absorbing complex. Ion exchange is not important for isotopes of the rare-earth elements, zirconium, ruthenium, and caesium. Rare-earth radionuclides usually form insoluble humates, phosphates, carbonates, and sulphates. However, because they have a lower solubility than the corresponding compounds of Ca, they tend to replace it. Isomorphic replacement is important for the absorption of ^{137}Cs, which replaces potassium in clay minerals.

The behaviour of radionuclides in soils can be substantially different from that of the macrocomponents—their isotopic and nonisotopic carriers (Ch. 2). The geochemistry of radionuclides is governed by their trace concentrations and forms of existence, which are distinct from the forms of their carriers in the formed soil. Thus, the concentration of ^{90}Sr from global fallout on the surface of a soil is on the average 10 mCi/km^2 or $n \times 10^{-14}$% in the surface layer of the soil [8.13]. This is approximately 10^{11} times less than the concentration of stable strontium in soils. Thus, radionuclides do not compete, so to speak, for the place on the absorbed surface, i.e. the absorption capacity for them is thus infinite.

The absorption rate of radionuclides by soils depends on their form and primarily on their solubility. Water-soluble forms of radionuclides react with soil absorbing complex more readily than do the insoluble forms.

The soluble fraction of the global fallouts is [8.13] for ^{89}Sr (70-90%); for ^{90}Sr (30-90%); for ^{137}Cs (30-80%); for ^{144}Ce (4-40%); for ^{106}Ru + ^{106}Ru (50-90%); for ^{95}Zr $^{+}$ ^{95}Nb (60-90%).

On the whole the soluble fraction increases in the series: ^{144}Ce, ^{91}Y < ^{95}Zr + ^{95}Nb < ^{137}Cs $^{<}$ ^{106}Ru + ^{106}Rh < ^{90}Sr. The charge of aerosol particles is important. Neutral and negatively charged particles in the fallout are bound less firmly to soil colloids than positive ones, as colloids have mainly negative charge of outer layer.

For the first few years after their fall to the soil surface radionuclides are concentrated in the upper layers and behave very differently from their natural stable isotopes (isotopic carriers) in the soil. It is only in the course of time that the forms in which radionuclides exist vary and they begin to enter the migration cycles characteristic of their isotopic carriers.

Depending on the soil composition the sorption of radionuclides will change, which can also be said of their bond to the soil, as well as the mechanisms underlying the process of sorption. For the same radionuclide, absorption by soil will increase in the series: soddy podzolic-sandy loam < soddy podzolic-loam < black soil varieties. The sequence is due to the increase in the series of the absorption capacity, which is due to the humus content, silt fraction, montmorillonite minerals in the clay minerals, and the formation of stable poorly soluble humates. The composition of gel films is also significant. The radionuclides absorbed by the soil grow in concentration with the number of particles containing Fe, Mn, and Al oxides and organic matter, with which radionuclides form poorly soluble compounds. The behaviour of radionuclides in soils is illustrated by data for ^{90}Sr, which has been studied more thoroughly than other radionuclides [8.13]. ^{90}Sr is carried away from soils in humid climates more easily because of the abundance of soluble organic compounds. In arid climate soils, ^{90}Sr accumulates at evaporation barriers with the formation of poorly soluble carbonates. In the conjugate landscapes ^{90}Sr is accumulated in lower elements of the relief. It is least present in watershed soils with washing out conditions.

The radioactive pollution of the environment by the transuranium radionuclides, or the isotopes of Np, Pu, Am, and Cm is of specific nature. They have the longest half-lives (up to tens of thousands of years) and are thus the sources of practically continuous contamination for humanity. The transuranics do not have any stable analogues in nature, in the same way as the isotopic and non-isotopic carriers of fission products, which affects their geochemical properties [8.14].

Plutonium is predominantly immobile in soils and sediments. Even when it enters the soil as a soluble compound it is mainly hydrolyzed. Almost all the relatively soluble Pu that can migrate is the form of $Pu(OH)_n$ in the soils. It is assumed that Pu which can be assimilated by plants can also exist as bicarbonate and carbonate complexes, or complexes with low molecular organic ligands. It has been established that in arid climates Pu can penetrate the soil to a depth of 30 cm. On the whole the behaviour of Pu(IV) has very much in common with Th(IV).

Hydrolysis also determines the behaviour of Am and Cm in soils. However, $Am(OH)_3$ and $Cm(OH)_3$ are more soluble and accessible for the plants than $Pu(OH)_4$.

The behaviour of Np in soils has not been well studied. Its compounds are most soluble among the other transuranium elements and are most accessible for biota.

The migration of radionuclides from places of temporary or final burial has not been adequately studied. On the whole, the behaviour of each radionuclide in the soils is determined by the same circumstances that have been described. However, there are also peculiarities which distinguish the behaviour of radionuclides that enter the soil in the form of aerosols of fallouts or with liquid waste. In the first case radionuclides (in the form of submicron particles) get into the natural medium and enter the geochemical cycles. In the second case the soil receives vast quantities of compounds that markedly change the chemical and physical chemistry of the medium. The migration of radionuclides in the latter case will be governed by the situation that has arisen as a result of the interaction between the contaminants and nature. To illustrate, there is the fairly widespread method of burying liquid waste with average activity in trenches filled with lumps of weathered limestone impregnated with alkali. The trenches were covered with a dense layer of soil [8.15]. This creates an alkaline buffer hindering the migration of ^{90}Sr and ^{137}Cs. However, in this environment ^{60}Co was found to be quite mobile and readily migrated with the ground waters beyond the storage. Co, which has two oxidation numbers, was in the form of Co^{2+}, which is most soluble and stable at pH = 9.5. Moreover it formed anion complexes with inorganic and organic ligands, namely, the low molecular soil fulvoacids and EDTA of contaminating solutions.

^{99}Tc has seven oxidation numbers. In the oxidizing conditions of the storage Tc(VII) is most stable in the form TcO^-_4, which is soluble in ground water and is mobile in soils. Tc can be reduced to Tc(IV) on reduction barriers, such as organic matter, thus giving rise to the weakly soluble TcO_2. In these conditions uranium is also mobile and migrates in the ground water as a carbonate complex. For radionuclides forming decay chains, it is necessary to take into account the mobility of intermediate products. The migration of mobile parent radionuclides can result in pollution by less mobile daughter nuclides. For example, there is the progressive pollution by ^{238}Pu and ^{240}Pu, which accumulate from the more mobile ^{242}Cm (162 days) and ^{244}Cm (18 years).

8.4 The Pollution of the Environment by Natural Radionuclides

Anthropogenic pollution by natural radionuclides (^{232}Th, ^{238}U, ^{235}U and their decay products, as well as by ^{40}K) occurs in extraction and processing of minerals from deep deposits. This results in localities in the biosphere where concentrations of uranium, thorium, radium, radon, polonium, and other natural radionuclides substantially exceed the natural background. Pollutions of this kind are considered less active. However, they contain a number of long-lived radionuclides (^{232}Th, ^{230}Th, ^{238}U, ^{234}U, ^{226}Ra, ^{231}Pa) and will, therefore, exist practically for ever. This anomaly will always be a permanent source of high-energy short-lived radionuclides, the greatest hazard being the isotopes of radon (Ch. 5).

The potential sources of pollution by natural radionuclides are enterprises at the initial stage of the nuclear fuel cycle, the production and use of agricultural fertilizers, the use of building materials, coal-fired power stations, and the recovery of hydrocarbon fuel [8.16-8.18].

8.4.1 Pollution of the Environment by Natural Radionuclides at the Initial Stage of the Nuclear Fuel Cycle

Contamination Sources

The initial stage in the nuclear fuel cycle includes enterprises that recover and process uranium ore. The ore is extracted from mines, by open-cast mining, or underground leaching. In the first two cases the benefication process involves grinding the ore and extracting the uranium by leaching. The reagents are H_2SO_4, sometimes Na_2CO_3 with the addition of oxidants (H_2O_2, NaOCl, etc.). Industrial grade ore contains 0.1-0.2% of uranium. Thus, solid wastes after leaching make up 99% of the original mass of ore and contain most of the decay products of uranium. The overall activity of one tonne of ore containing 0.2% U (2 kg U) is approximately 7.9 mCi (2.9×10^8 Bq)

[8.1]. In the case of sulphate leaching up to 50% of ^{230}Th can be transferred (and, presumably, ^{234}Th) and less than 1% of ^{226}Ra. After the extraction of uranium from the mother liquor about 5% will be thorium and 0.2% will be radium. All in all the uranium will have about 14% of the activity of the mother liquor [8.2]. Underground leaching, a method in which a leaching solution (H_2SO_4) is pumped into an injection well and pumped out from a working well, is only used when the geological conditions are favourable. As before, uranium is recovered from the sulphate solution by ionic exchange, extraction, and precipitation.

Primary ore processing plants produce the following sources of environmental pollution: gaseous ^{222}Rn from the ore and waste; liquid waste (solutions after the extraction of uranium, mine water, etc.), solid waste after leaching, hydrate sediments, and pulp. Solid waste and pulp (tails) are placed in open tailing dumps.

After drying, solid pulp particles can undergo wind abrasion and be removed beyond the tailing dumps. To prevent this, the solid waste is covered with rock or polymers, bitumens, or other binding reagents, and then the site recultivated. To remove radionuclides from liquid waste, in particular from ^{226}Ra and ^{230}Th, the solution is neutralized with lime, Ba salts, evaporating or filtering ponds, sorption on zeolites, or pumped into deep wells.

The Migration of Radionuclides

The worst ecological and geochemical problem is the migration of radionuclides away from where the waste products are buried. Migration from the tailing dump of solid waste takes place due to wind, climatic precipitation and vegetation [8.17].

Deflation can only occur when the integrity of the deactivation cover is violated. It can be avoided if care is taken to maintain it in reasonable time. Deflation is worst in arid climates. In humid climates water erosion of waste and the migration of radionuclides with atmospheric precipitation, which filters through the waste are the main causes. The composition of waters also changes. During filtration the water becomes richer in many of the relatively soluble components in the waste. Depending on the technology the concentration in the ground water of anions (SO^{2-}_4, Cl^-, HCO^-_3, NO^-_3, etc.) and cations (K^+, Na^+, Mg^{2+}, etc.) rises rapidly. This results in a significant increase in the chemical corrosiveness of the ground water in the waste, which intensifies the transfer of radionuclides into the water. Experiments on the repeated leaching of waste by a 0.1 N solution of HCl, modelling natural processes, have shown that ore waste can yield up to 20% of ^{238}U and over 25% of ^{226}Ra.

To study the migration of radionuclides and assess its intensity, particularly at the start of the process, when the concentration of the elements transferred beyond the tailing dump do not differ from the natural background, the non-equilibrium isotopic method should be used. The isotopic ratios $^{234}U/^{238}U$, $^{230}Th/^{232}Th$, and $^{228}Th/^{232}Th$ are most informative. The first two can serve as pollution tracers, while the latter are indicators of the migration intensity.

The research can be done with soil and water. The background activity of $^{234}U/^{238}U$ in surface and ground waters, as well as in the humus horizon of soils usually exceeds the equilibrium value (>1). By contrast, the typical value for uranium ores is very close to one. The removal of uranium from tailing dumps with runoff either in a dissolved state or as a suspension will show up as a drop in the $^{234}U/^{238}U$ ratio.

The $^{230}Th/^{232}Th$ ratio in uncontaminated soils (the background) does not usually exceed 1.0 ± 0.2 [8.17]. Uranium ores usually contain uranium and its decay products (including ^{230}Th). ^{232}Th and its decay products exist in these ores in smaller concentrations, whence we have $^{230}Th/^{232}Th$ >> 1.

The migration of uranium can be retarded by a reduction barrier (e.g., peat or gley soils). The ^{230}Th will in these conditions migrate in the form of soluble

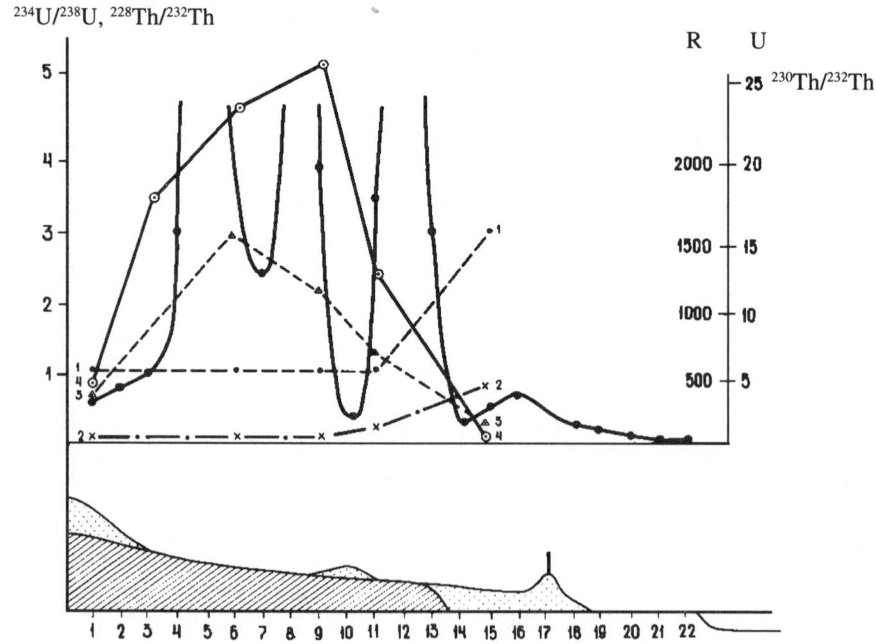

Fig. 8.4. Profiles across the tailing dumps of ore wastes: 1—wastes; 2—deactivation cover; $^{234}U/^{238}U$, $^{228}Th/^{232}Th$, and $^{230}Th/^{232}Th$ are the activity ratios of isotopes, U is the uranium concentration, pCi/g, R is gamma activity, μR/h

organic complexes which will show up as a $^{230}Th/^{232}Th$ ratio exceeding the background.

The isotopic ratio $^{228}Th/^{232}Th$ is a sensitive indicator of leaching (Ch. 5). When ore waste is leached by ground waters formed by atmospheric precipitation reacting with tails, this ratio becomes smaller (< 0.5) than the background for the soils, viz., about 0.8-1.0 (Fig. 8.4).

8.4.2 Pollution of the Environment by Natural Radionuclides from Coal-Fired Power Stations

Power stations of this type are the main consumers of coal. To produce one MW of electricity it is necessary to burn approximately 3×10^3 tonnes of coal a year. The ash produced by coal is 5-10%. Hence, to produce 1 MW per year of electric power, 300 tonnes of solid wastes are formed on average. If we take into account that the annual world consumption of coal is approaching 3×10^9 tonnes, we have a fairly impressive figure for the mass of waste. Most of the ash and cinder formed when the coal is burnt is dumped. However, a certain fraction escapes as aerosol particles, or volatile ash through the ash-collectors into the atmosphere. The cleansing efficiency for most contemporary thermal electric power stations in Russia is 98.5%. This indicates an average discharge into the atmosphere of 4.5 tonnes of aerosol particles per 1 MW per year, though this figure may actually be much higher.

When the coal is burnt volatile elements can be concentrated in the ash dump, depending on the volatility of the oxides. The least volatile elements will practically all be accumulated in solid waste. The more volatile will escape into the atmosphere and when cooled will either condense on the aerosols, or form independent aerosol particles. Balance data for coal-fired station have shown that the gas phase and aerosol particles containing no less than 75% of the uranium, 63% of the thorium and 50% of the radium in the coal were discharged into the atmosphere. In addition the ash is 1.2 times richer than the original coal in uranium, 2.2 times in thorium, and nearly 5.0 times in radium [8.17].

Thus, two ways of pollution are possible, i.e. at the expense of aerosol fallout and as a result of the utilization of the ash.

The relatively low activities of the solid waste from coal-fired power stations do not markedly differ from those of natural soils. However, taking into consideration the great volumes of this waste and that the form in which it exists is different from soils, we would expect, in the course of time, an increase in the radiation background (Fig. 5.12). The solubility of coal ash in 1 N HCl, which characterizes the biological availability, reaches 40%. Estimates of the pollution of the environment by natural radionuclides from a typical thermoelectric power station are given in Table 8.5. The calculations were based on a power station operating for 20 years. Data for the neighbourhood of a relatively old coal-fired power station with an old system of gas purification, show that 150 km away the content of radionuclides on the surface of a glacier is 50 times greater [8.2]. One more example. In a 5-10 km neighbourhood of a thermoelectric power station burning brown coal, the concentration of radionuclides on the surface of soil is 1.5-3.0 times greater than the background level.

Table 8.5 Discharge of Natural Radionuclides, Accumulation in Soil and Content in the Atmosphere in the Neighbourhood of a 1 GW Coal-Fired Power-Station Region [8.3]

Parameter	^{226}Ra	^{228}Ra	^{210}Pb	^{210}Po	^{232}Th	^{40}K
Annual discharge, Ci	0.53	0.3	2.2	2.0	0.53	5.3
Accumulation in soil, Ci	10.5	2.5	31	19.0	–	105
Density of on-site pollution, mCi/km^2	10.5	2.5	31	19.0	–	105
Concentration in the air, 10^{-8} Ci/l	1.7	1.1	4.0	3.9	1.7	–

In many countries, e.g. Poland, ash from brown and anthracite coals is used for calcification and to improve the structure of arable soils. Ash is also regarded as a fertilizer, and the ash from hard and brown coals is used to produce cement, bricks, and other building materials. In all cases, particularly in agriculture, it is necessary to assess the radiation properties of the utilized waste [8.16].

8.4.3 Pollution of the Environment by Natural Radionuclides as a Result of the Production and Use of Agricultural Fertilizers

A rise in the radiation background of arable lands at the expense of mineral fertilizers is particularly significant for humankind since in this way radionuclides enter the human organism with agricultural products . This process is intensified by the chemization of agriculture. Potassium fertilizers are the source of ^{40}K that enters the soil. However, the additional activity does not usually go beyond the limits of fluctuation of the natural background. The use of mining waste has not been studied. The use of ash from coal-fired electric power stations for soil liming is an example. This waste utilization must be conducted with great care and thoroughly studied beforehand.

On the other hand, the use of phosphate fertilizers has been given all the attention it deserves [8.16].

Natural phosphorites are used to produce phosphorous fertilizers and have higher concentration of radionuclides in the uranium series. The concentration of uranium in phosphorites sometimes reaches industrial values and it is extracted from the ore as a residual product. The radioactivity in phosphorites deposits is as high as 5 Bq/g for ^{238}U, and 0.16 Bq/g for ^{232}Th.

The redistribution of uranium and its decay products during the processing of the phosphate ore depends on the process flow sheet. When the treatment is wet, the phosphate ore is processed with sulphuric acid. The end product is phosphoric acid, which dissolves the uranium, a considerable fraction of the thorium, and the gypsum, leaving only radium.

When the phosphate ore is processed by dry annealing with coke and silicate most of the ^{226}Ra remains in the cinder together with the calcium silicate, while the uranium dissolves in the phosphoric and phosphorus acid.

Fertilizers obtained from phosphoric acid (e.g. ammonium phosphate) contain uranium and thorium, though they are free of radium. Triple superphosphate is obtained from phosphoric acid and commercial ore. That is why it has a high concentration of uranium and thorium (due to phosphoric acid) and radium from the decay products (due to the ore). Phosphate flour obtained by grinding the ore does not differ from the latter with respect to radionuclides. The concentration of U-series radionuclides in phosphoric fertilizers around the world fluctuates within the limits of 70-2400 Bq/kg. In Russia it is at a maximum in phosphate flour [8.4].

Thus, the continuous use of phosphorous fertilizers increases the concentration of natural radionuclides in arable soils. In the USA the ^{238}U concentration in arable soils is double that in virgin lands. The concentration of natural nuclides is 10-30% higher in Taiwan in fertilized soils in rice fields than in those where fertilizers are not used [8.4].

Depending on the type of fertilizer, i.e. on the processing technology used for the raw materials, radionuclides can enter the soil in forms that are different as far as their mobility is concerned. Thus, the fraction of mobile forms of natural radionuclides in phosphate flour is almost the same as that in soils that were not fertilized.

By contrast the fraction of mobile forms in the atmosphere is higher. Uranium has the maximum mobility while thorium has least mobility. Increasing the contact of fertilizers with the soil substantially changes the forms of radionuclides and increases the mobility (forms accessible to plants) of ^{238}U, ^{226}Ra, ^{210}Pb, ^{210}Po.

Hence, the number of water-soluble forms of ^{238}U is 10 times greater in fertilized soils, while the acid-soluble forms are four times higher than in the original soil [8.16].

8.4.4 Pollution of the Environment with Natural Radionuclides when Mining Hydrocarbon Raw Materials

Radionuclide pollution from recovering hydrocarbon raw materials (oil and gas) is due to the brines, particularly from water-oil or water-gas interfaces. They contain 100-1000 times higher concentrations of radium than the background values. Other daughter products in the radioactive series also exceed the background values by 100 times. The elements ^{226}Ra, ^{228}Ra and ^{230}Th are very significant due to long-lived radioactive anomalies. The radioactivity of boring brine can lead to the pollution of wells and the surface of soil around them (Ch.5 [5.3]).

When the soil becomes contaminated due to the discharge of boring brine, stable radiochemical anomalies are formed in which the behaviour of individual radionuclides obeys rigid regularities, which at first glance, seems paradoxical. Thus, the mobility series is $^{226}Ra = {}^{228}Ra < {}^{230}Th < {}^{228}Th < {}^{224}Ra$. This must be because:

1. Relatively long-lived ^{226}Ra, ^{228}Ra and ^{230}Th must have contaminated the brine. ^{226}Ra and ^{228}Ra are less mobile in polluted soils than are the isotopes of thorium. The paradox is due to the ion-exchange mechanism involving the influx of radium into soil colloids. In excess $CaCl_2$ organic and mineral colloidal particles form stable aggregates 'enclosing' the radium ions [8.17]. Owing to their high charge, the ions of $^{230}Th^{4+}$ do not tend to exchange in soil, absorbing complexes. This means that they are less firmly bound and penetrate to a substantially greater depth than $^{226}Ra^{2+}$ and $^{228}Ra^{2+}$.

2. The more short-lived ^{228}Th and ^{224}Ra decay in a brief period of time. In a month all the ^{224}Ra atoms that come with the brine will have decayed and will give place to other atoms that have accumulated in the already contaminated soil in the form of recoil atoms as a result of the radioactive decay. A complete exchange of ^{228}Th atoms will actually take place in twenty years. Depending on the age of the anthropogenic anomaly, some of the ^{228}Th atoms are primary, arriving during the pollution, while the rest are secondary, accumulating later. If the contamination took place 10 years ago or more, only 3% of the primary ^{228}Th remains. The recoil atoms of ^{228}Th and ^{224}Ra will behave in exactly the same way as in uncontaminated soils, where the mobility of radium is higher than that of thorium.

The ^{226}Ra in boring brine may be the source of the ^{222}Rn in natural gas. The concentration of radon in the natural gas that is delivered to the consumer will depend on the length of time it took to transport. In 38 days its concentration falls by approximately 1000 times. The main pollution hazard emerges at the gas plants in the proximity of operating wells. The contamination of the inner surface of gas transmission pumps and pipes by ^{210}Pb and ^{210}Po or the products of radon decay.

8.4.5 The Pollution of the Environment by Building Materials

Building materials are invariably the products of nature (granite, marble, tuff, gypsum) and do not, as a rule, differ by radioactivity from the environment. Radon concentration is possible in poorly ventilated buildings that are plated inside with materials that have high abundance ratios of uranium and thorium (e.g. granites). Artificial building materials that are obtained from the wastes of mining industry deserve more attention and study, e.g. phosphogypsums from phosphates, blast furnace cinder, waste products from the aluminium industry, etc. [8.16, 8.21-8.23]. The aerial supply of radon isotopes into the human organism in premisses plated inside with such materials is an important ecological problem.

REFERENCES

8.1. Aleksakhin, P.M., 1983. Nuclear Energetics and Biosphere. Energoizdat, Moscow.
8.2. The Sources and Action of Ionizing Radiation. UNO Scientific Committee of Atomic Radiation Action, 1978. New York, **1**.
8.3. Nuclear Energetics, Man and the Environment, 1984. (Ed. A.P. Aleksandrov). Energoatomizdat, Moscow.
8.4. The Ionizing Radiation: the Sources and Action, 1982.—Rep. of UNO Scientific Committee of Atomic Radiation Action, in two volumes. New York, **1**.
8.5. The Sources and Action of Ionizing Radiation, 1978.—Rep. of UNO Scientific Committee of Atomic Radiation Action. New York, **1**.
8.6. The Irradiation from the Natural Sources of Ionizing Radiation, 1984.—Rep. of UNO Scientific Committee of Atomic Radiation Action. New York.
8.7. Gromov, B.B., Spitsyn, V.I., 1975. Artificial Radionuclides in Sea Water. Atomizdat, Moscow.
8.8. Kogan, P.M., Nazarov, I.M., Fridman, Sh.D., 1976. The Principles of Gamma-spectrometry of Natural Media. Atomizdat, Moscow.
8.9. Izrael, Yu.A., Petrov, V.I. Avdonm, S.I. et al., 1987. Radioactive Pollution of Natural Media in the Zone of Chernobyl Nuclear Power Station.—Meteorology and Hydrogeology, No 2, 5-18.

8.10. Izrael, Yu.A., Sokolovskii, V.G., Sokolov, V.B. et al., 1988. Ecological Consequences of Radioactive Pollution of Natural Media in the Region of Chernobyl Nuclear Power Station.—Atomic Energy, **64**.

8.11. The Information about Chernobyl NPS Accident and Its Consequences for IAGAE, 1986.— Atomic Energy, **61**, No 5, 301-320.

8.12. Evolution of Atmospheric Dispersion Model Applied to the Release from Chernobyl, 1990. Wien, Osterrech.—Betr. Meteorolog. Geophys., 329, 197.

8.13. Pavlotskaya, F.I., 1974. The Migration of Radioactive Products of Global Fallouts in Soils. Atomizdat, Moscow.

8.14. Transuranic Elements in the Environment, 1980. (Ed. W.C. Hanson). Technical Information Center/U.S. Department of Energy.

8.15. Olsen, C.R., Lovry, P.D., Kee, S.Y. et al., 1986. Geochemical and Environmental Processes Affecting Radionuclides Migration from a Formerly Used Seepage Trench.—Geochemica et Cosmochem. Acta, **50**, No 4, 593-608.

8.16. Heavy Natural Radionuclides in the Biosphere.—Migration and Biological Action in Population and Biocoenosis, 1990. (Ed. R.M. Aleksakhin). Nauka, Moscow.

8.17. Titayeva, N.A., 1991. Geochemistry of Natural Radionuclides in Hypergenesis Zone.—The Problems of Radiochemistry and Cosmology. Nauka, Moscow, 64-129.

8.18. Chambers, B.D., Cassaday, V.J., Low, L.H., 1989. Exposures from Mining and Mine Tailings.—Radiat. Phys. and Chem., **2**, 295-307.

8.19. Titayeva, N.A., Aleksakhin, R.M., Taskaev, A.I., and Maslov, V.I., 1980. Migration of Heavy Natural Radionuclides in a Humid Climatic Zone: Nature Radiation Environment III. Proceeding of a Symposium, held at Houston, Texas, April 23-28, 1978. (Ed. Thomas F. Gesel, Wayne M. Lomder), **1**. Published by Technical Information Center/US Department of Energy.

8.20. Titayeva, N.A., Maslov, V.I., Taskaev, A.I., and Aleksakhin, R.M., 1983. Radioactive Elements in Anthropogenic Pollution of the Environment.—The Probelms of Radiogeology. Nauka, Moscow, 224-237.

8.21. Crameri, R., Brunner, N.N., Buchli, R. et al., 1989. Indoor Rn Levels in Different Geological Areas in Switzerland.—Health. Phys., **1**, 29-38.

8.22. Yorsek, L., Humar, M., Ilic, R. et al., 1990. Indoor and Outdoor Radon Measurements in the Saleska Valley, Nucl. Track. Radiat. Meas., **17**, No 1, 73.

8.23. Krisyuk, E.M., 1989. Radiation Background in Buildings. Energoizdat, Moscow.

9 The Influence of Radioactive Radiation on the Geological Medium

'Radiochemical changes presumably play a great role in the structure of the earth's crust. Of particular importance is the process connected with the decomposition of water molecule which penetrates the earth's crust and the whole of its solid matter throughout.'

V.I. Vernadsky

9.1 The Interaction Between Radiation and Matter

When radioactive radiation passes through a substance the particles or photons interact with the atoms. The principal parameters in this interaction are the mass, charge, energy (for the particles) and the atomic number, density, and mean ionization potential (for the substance).

Heavy charged particles (*alpha particles, protons, deuterons, etc.*) are mainly retarded when passing through the substance due to their interaction with atomic electronic shells. This is accompanied by the ionization and excitation of the atoms and molecules. The loss of energy per unit length of free path $\left(-\dfrac{dE}{dx}\right)$ is called linear energy transfer (LET). For the heavy particles it is [9.1]

$$-\frac{dE}{dx}=\frac{4\pi Z_e^2 e^4 \mathrm{N_A} Z_\rho}{mv^2 A_s}-\left[\ln\frac{2mv^2}{\varphi\left(1-\beta^2\right)}-\beta^2-\frac{\delta}{2}\right] \qquad (9.1)$$

where e is the charge on an electron; Z_e is the charge of the moving particle, N_A is Avogadro's number, Z is the effective atomic number of the absorbing substance; m is the electron rest mass; v is the velocity of the particle, A_s is the mean atomic mass of the substance, φ is the mean potential of the excitation of the atoms in the substance; β is v/c (c is the light speed); $\frac{\delta}{2}$ is the ionization effect correction; ρ is the density of substance.

It follows from (9.1) that the LET and consequently the ionization density increase by the end of the free path.

The electrons of ionizing radiation (β *particles*) lose their energy as a result of inelastic scattering by electrons and atomic nuclei in the absorbing material. This leads to ionization of the atoms and causes Bremsstrahlung. The loss of energy per unit length (LET) for a fast electron is

$$-\frac{dE}{dX}=\frac{2\pi e^4 N_A \rho Z}{mv^2 A_s}-\left[\ln\frac{mv^2\times E}{2\varphi\left(1-\beta^2\right)}\right.$$
$$-\left(2\sqrt{1-\beta^2}-1-\beta^2\right)\ln 2$$
$$\left.+1-\beta^2+\frac{1}{8}\left(1-\sqrt{1-\beta^2}\right)^2-\delta\right] \tag{9.2}$$

Thus, the braking power of a substance is proportional to the density of medium, its effective atomic number and the number of atoms per cm^2. For particles of different types it is directly proportional to the square of the charge of the ionizing particle and inversely proportional to its kinetic energy.

The path covered by the particle in a substance is called the track. The number of ions per unit track of a heavy charged particle is considerably greater than that in the track of an electron.

The gamma and X-ray photons lose their energy as a result of photoelectric absorption and the formation of electron proton pairs (Ch. 1). The electrons that emerge lose their energy due to ionization. The absorbed energy of gamma or X-ray E_{abs} is defined as

$$E_{abs}=E_0\left(1-e^{-\frac{\mu a}{\rho}d}\right) \tag{9.3}$$

where E_0 is the initial energy, μ/ρ is the mass absorption coefficient, d is the thickness of the absorbing layer, g/cm^2, ρ is the density of the material g/cm^2.

Neutrons only interact with atomic nuclei, thus causing elastic and inelastic scattering, nuclear reactions and nuclear fission. The radiation changes in the substance depend on the absorbed dose of radiation.

Recoil nuclei. Besides α, β, and γ-radiation, significant radiation effect can be produced by the recoil nuclei, which can be defined as the remains of an atom that emerged as a result of radioactive decay. The energy of radioactive decay is partially transmitted to the recoil atom and is then transformed into kinetic energy (E_{at}). The redistribution of energy between the recoil atom and the particle can be estimated from

$$E_{at} = \frac{M_{part}}{M_{at}} \varepsilon_{part} \tag{9.4}$$

where E_{at} and ε_{part} are the energies of the recoil atom and the particle, respectively, and M_{at} and M_{part} are masses of the recoil atom and particle.

Example.

1. ^{222}Rn decay, viz., $^{222}\mathrm{Rn} \xrightarrow{\alpha} {}^{218}\mathrm{Po}$ yields recoil nuclei ^{218}Po and α-particles with energy 5.482 MeV. The energy of the recoil nuclei for ^{218}Po is

$$E_{at} = \frac{4}{218} \times 5.482 = 0.1\,\mathrm{MeV}$$

The recoil energy of the alpha decay is of the order of tenths of MeV. For β decay the kinetic energy of the recoil atom varies from hundredths to tens of eV. During the emission of a γ-photon, the recoil atom acquires a kinetic energy of the order of tens of eV.

The atomic binding energy in the molecule is approximately 2 to 5 eV. Therefore, in the case of alpha decay the chemical bonds in the molecule containing the parent atom are disrupted.

The free recoil atom begins to move in the medium at a great speed, thus becoming a 'hot atom'. For β-decay, K-capture, and isomer transition, no chemical bonds are broken.

A hot atom (with an energy greater than thermal one) interacts with the medium causing the ionization and the excitation of molecules. In solids it traverses a path of 100-1000 Å, experiencing inelastic atom-atom collisions. At the end of the path (~10 Å) a rigid sphere collision occurs. The energy of recoil atom is expended to create cascades of displaced atoms and heating the disruption zones, called 'hot zones'. The time within which the hot zone is formed is of the order of 10^{-12} s, and the temperature in it can reach 10^5 K (see Ch. 5, [5.7]).

Cooled to thermal velocities, the recoil atoms can enter ordinary chemical reactions and isotopic exchange (secondary confinement). The region of the stabilization of recoil atoms is metastable, but with an increased energy and defects in the crystal lattice. External factors (heating, irradiation) can cause a reconstructing of the primary molecules, which is called annealing.

9.2 Chemical Transformations of Substance Under Radiation

According to modern notions, the interaction between any type of ionizing radiation and matter reduces to two basic primary processes, i.e. ionization and the excitation of molecules and atoms [9.1]:

$$M \rightsquigarrow M^+ + e^- \quad \text{(ionization)}$$

$$M \rightsquigarrow M^* \quad \text{(excitation)}$$

These primary processes depend on the absorbed radiation dose and are practically independent of the substance. On average, for each 100 eV of absorbed energy approximately 4 pairs of ions and 0.5-1 excited molecule is produced.

The positive ions formed by the radiation, as well as the excited molecules and electrons with thermal velocities take part in secondary chemical reactions, which depend on the LET, the dose rate, the chemistry, and physical chemistry of the medium.

The unstable molecular ions decay into a free radical and an ion:

$$M^+ \rightarrow R_1 + R_2^+$$

The positively charged ions enter an ion-molecular reaction with surrounding molecules, which results in the formation of radicals and protonated molecules

$$M_1^+ + M_2 \rightarrow R + M_1H^+$$

After the energy of the electron has decreased to values below the ionization potential and molecular excitation, the electrons can

(a) be captured by the molecules having affinity to the electron, sometimes with a subsequent dissociation

$$M + e^- \rightarrow M^-$$

$$MX + e^- \rightarrow M + X^-$$

(b) form a solvated (e_s^-) electron in the solution or a captured (e_t^-) electron in a solid:

$$e^- + nM \rightarrow e_s^-(e_t^-)$$

The negative ions recombine with the positive ones to form molecules:

$$M_1^+ + M_2^- \rightarrow \begin{cases} M_1M_2 \\ M_1 + M_2 \end{cases}$$

The excited molecules disintegrate into stable molecules or radicals

$$M* \rightarrow \begin{cases} M_1 + M_2 \\ R_1 + M_2 \end{cases}$$

The active radicals, ions, molecules, and atoms formed as a result of these processes react in turn with surrounding medium. All the above processes are more distinct in the gas phase. In condensed phase recombination occurs.

9.2.1 Radiolysis of Water

Ionizing radiations cause water radiolysis. The maximum amount of radiolysis products is formed by radiations with a high ionization density, e.g. α-particles.

As a result of the radiolysis of water the following processes take place.

1. Ionization and excitation of water molecules, i.e.,

$$H_2O \rightsquigarrow H_2O^+ + e^-$$
$$H_2O \rightsquigarrow H_2O^*$$

2. The solvation of electrons and the subsequent reaction with water molecules, i.e.,

$$e_{aq}^- + H_2O \rightarrow \dot{H} + OH^-$$

3. The formation of $\dot{H}$ and $\dot{O}H$ radicals as a result of the reactions of ions and excited water molecules, i.e.,

$$H_2O* \rightarrow \dot{H} + \dot{O}H$$
$$H_2O^+ + H_2O \rightarrow H_3O^+ + \dot{O}H$$

4. Radicals recombine where their concentration is high, i.e.,

$$\dot{H} + \dot{H} \rightarrow H_2$$

$$\dot{H} + \dot{O}H \rightarrow H_2O$$

$$\dot{O}H + \dot{O}H \rightarrow H_2O_2$$

5. The formation of free oxygen for radiation with high LET:

$$H_2O_2 + \dot{O}H \rightarrow H\dot{O}_2 + H_2O$$

$$H\dot{O}_2 + H\dot{O}_2 \rightarrow H_2O_2 + O_2$$

In dilute aqueous solutions only an infinitesimal fraction of solute is affected by radiation. The changes are mainly due to reactions with water radiolysis products. This is accompanied by redox reactions. Water radiolysis products may be both oxidizing and reducing agents depending on the acidity of the solution and the character of the solute. A solvated electron is a reducing agent and in acid solutions it reduces hydrogen ions:

$$H_3O^+ + e_s^- \rightarrow H + H_2O$$

Hydrogen atoms usually have reducing properties, which are intensified with increasing pH.

In acids hydrogen atoms are oxidizers since in a reaction $H + H^+ \rightarrow H_2^+$ ion is formed, which functions as the oxidizer, $\dot{O}H$ radicals are usually oxidizers. However, in the presence of strong oxidizers (e.g. $KMnO_4$) they can become reducing agents. For pH > 9 electrolytic dissociation of the $\dot{O}H$ radical occurs to form $\dot{O}^-$, which is a weak reducing agent $\dot{O}H \rightarrow \dot{O}^-H^+$.

The direct radiolysis of a solute only occurs in concentrated solutions.

9.2.2 The Radiolysis of Organic Matter

The radiolysis of organic matters leads to cleavage of the C—C, C—H, and C-function bonds. As a result of the radiolysis a large number of radicals appear which interact with one another. An inevitable product of the radiolysis of organic compounds is hydrogen, i.e.,

$$RH \rightsquigarrow \dot{H} + R; \quad RH + \dot{H} \rightsquigarrow R + \dot{H} \text{ (radical } H_2\text{)}$$

$$RH \rightsquigarrow H_2 + \text{unsaturated organic compounds (molecular } H_2\text{)}$$

The second important product is methane, which is formed from the terminal groups:

$$\dot{C}H_3 + RH \rightarrow CH_4 + \dot{R}$$

In terms of radiation stability, hydrocarbons form the following series: aromatic > saturated > unsaturated. The action of ionizing radiation on the mixture of organic compounds may cause the synthesis of new compounds, polymerization, etc.

9.2.3 Radiation Transformations in Solids

Ionizing radiation on solids also results in the ionization and excitation of molecules. However, the radiation yield in solids is substantially lower since the radiation energy absorbed by one molecule is rapidly redistributed among neighbouring molecules, and the effect is lower. The transmission of energy by elastic collisions increases the recombination probability.

In crystals deformations in the crystal structure can occur: the formation of vacancies, inclusions, dislocations of various kind, and dislocation wedges. As a result of the transfer of kinetic energy of the moving particle to the crystal atoms, the local area around its track is quickly heated up, as many as 6000 atoms being heated. The temperature can reach 4000 K, and within 10^{-12} s the area is in a melted state. The solidification process is accompanied by the emergence of a zone of mechanical stress, which leads to the formation of dislocations. In the melting zone the chemical and physical properties of the crystal are changed. In particular, the rate of dissolution in the track becomes higher.

Radiation causes changes in the surface of solids: active centres of defects and excited atoms appear on the surface. The specific surface of the solid phase may also increase. The irradiation of solids can result in the emergence of charges on their surfaces. For example, the surface of a dielectric irradiated by gamma rays becomes positively charged, while the excess electrons are redistributed within the solid. These changes lead to changes in the physical and chemical properties of solids: the rate of dissolution increases, the boiling point drops, the adsorbing power as well as the ability to form colloids grows.

9.3 Paleodosimetry

All these radiation effects in minerals depend on the absorbed dose of the radiation energy. The effects can remain in a mineral for hundreds of thousands or even million years giving information about the total dose accumulated by the mineral during its existence or during the time it has been irradiated. If the relation dose vs effect is proportional, then the corresponding radiation changes can be used to determine the historical absorbed dose (paleodose). Methods using radiation effects in minerals to evaluate a paleodose are called *paleodosimetric* and the corresponding minerals *paleodosimeters.*

Paleodosimetric methods are often used to age the mineral (or the duration of its last irradiation). All paleodosimetric methods essentially consist of the following steps:

(1) the magnitude of the radiation effect is determined in some way;

(2) the mineral is heated hot enough for the accumulated radiation changes to disappear and thus determine the primary state of the sample (zero-moment);

(3) the mineral is calibrated for determining the accumulated dose. To do this it is artificially irradiated by different doses of ionizing radiation and the magnitudes of the corresponding radiation effects are recorded. The obtained calibration curve in dose-effect coordinates is used to determine the value of paleodose D (from the intensity of the natural radiation effect measured earlier);

(4) the radioactivity A of the sample and/or its enclosing rock is determined;

(5) the sample's exposure time or radiological age is calculated from the relation:

$$t = f(D, A) \tag{9.5}$$

In principle, $t = \frac{D}{P}$, where P is the intensity of α-, β- and γ-radiation through the mineral.

The common drawbacks of paleodosimetric methods, which use various radiation effects, include:

(a) the inadequacy of artificial irradiation for calibrating to the natural irradiation of the mineral;

(b) the influence of secondary geochemical factors on the accumulation and preservation of the effect;

(c) the establishment of the zero-moment during the annealing process.

In this chapter the most widespread paleodosimetric methods are treated in greater detail.

9.4 The Radiation Electron-Hole Centres in Minerals

Natural minerals usually have point defects in their crystalline lattice. These include [9.2]: (1) vacant sites of anions and cations; (2) 'interstitial' atoms and molecules; (3) impurity ions (isomorphic and interstitial) as well as impurity radicals; (4) atoms in a state of disorder distribution; (5) aggregates of vacant sites and admixtures. Larger aggregates of defects give rise to clusters, and further to colloidal formations. Pores of different size are formed when vacant sites are accumulated.

The existence of defects in the structure of a mineral disturbs its electric neutrality. The ionizing radiation causes the defects to ionize. Some defects give up electron (this is described as the capture of a positive elementary charge-hole), while other defects capture the knocked out electron. This results in the formation of positively and negatively charged defects, which are called electron-hole centres. Examples of radiation centres in minerals are [9.2, 9.3].

1. In quartz impurity cations with a charge smaller than Si^{4+}, e.g. Al^{3+}, which replaces Si^{4+}, create a deficiency of a positive charge, and one electron in the shell of the oxygen anion becomes excessive. During irradiation the electron is knocked out from the oxygen anion O^{2-} nearest to Al^{3+}, which is then transformed to O^{-}. The absence of the electron in anion shell is designated as the capture of the hole, which behaves in the same way as a positively charged electron. This centre is called a hole centre.

2. The impurity cation with large charge produces a deficiency of negative charge, or an anion vacant site. It is charged positively and is a trap for electrons. For instance, Ni^{2+} impurity in LiF gives rise to Ni^{+} ions, which are electron centres. When an electron is captured by the vacant site of an anion electric neutrality is restored. The electron centre is formed, which is called an F-centre.

3. The vacant site of oxygen in the quartz structure becomes responsible for the excess of positive charge and functions as a trap for electrons (SiO_3^{3-} in silicic *o*-oxygen tetrahedron). A vacant site of this type captures the electron and forms the electronic centre in quartz, which is called an E_1-centre.

The radiation electron or hole centres that have thus emerged have an unpaired, and, consequently, paramagnetic electron. Therefore, centres of this kind are recorded by paramagnetic resonance.

For paleodosimetry E_1-centres in quartz are the most frequently used. They are preserved for 500 mln years and they have the advantage of being monotypical, since the E_1-centres are unpaired (single) electron occupying splitted Si—O bond. They are not associated with isomorphic impurities as Al-, Ti- and Ge-centres. There is also a functional dependence between the concentration of E_1-centres and the value of paleodose. It has been experimentally established that E_1-centres at the boundary of radioactive ores are mainly created by α-particles. At first α-particles form defects in the crystal structure in quartz (proto-centres), knocking out oxygen ions from the structural positions. Then the vacant sites capture the electrons that have emerged during ionization. In the case of background concentrations of radionuclides radiation centres are more often due to γ-radiation. In addition, defects that emerged during crystallization are filled with electrons and holes.

E_1-centres are often used to study ores [9.4].

To establish the age of uranium ore mineralization it is necessary to consider the concentration of E_1-centres in the quartz grains in contact with the uranium minerals in the ore.

The estimate comes from the equation

$$dC_E / dt = \sum f_j p_j(R) b_j C_n - KC_E^2 \tag{9.6}$$

where C_E is the local concentration of E_1-centres, f_j is the efficiency of the formation of E_1-centres due to the j-type radiation (α, β, γ), $P_j(R)$ is the flux of j-th radiation per dosimeter (quartz grain) of radius R from the concentration unit of the radiation source in the rock; b_j is the number of particles of j-th radiation, C_n is the concentration of emitter in the enclosing rock, KC_E is the rate of the spontaneous extinction of E_1-centres. The efficiency of formation of E_1-centres is determined experimentally. The numbers of E_1-centres in quartz formed by one particle or photon are [9.4] $f_\alpha = 1.02$; $f_\beta = 0.13$; $f_\gamma = 2 \times 10^{-4}$. The formulae for estimating age after the substitution of corresponding parameters are [9.4]:

$$t_\alpha = 2.79 \times 10^{-14}\left(C_E^\alpha - C_{E\,\text{backgr}}^\alpha\right) \cdot I_\alpha \tag{9.7}$$

$$t_\beta = 2.15 \times 10^{-13}\left(C_E^\beta - C_{E\,\text{backgr}}^\beta\right) \cdot I_\beta \tag{9.8}$$

$$t_\gamma = 3.94 \times 10^{-15}\left(C_E^\gamma - C_{E\,\text{backgr}}^\gamma\right) \cdot C_K \tag{9.9}$$

where I_j is the flux of corresponding radiation.

The average age will be $t = (t_\alpha + t_\beta + t_\gamma)/3$. As a result of the insignificant contribution of γ-radiation to the paleodose the last term can be neglected.

The rate of migration V of a uranium ore body, which has the form of a roll, was estimated on the basis of the relation [9.4]

$$V = \Delta I P_j^t f_j / 2\left(C_E^t - C_E^o\right) \tag{9.10}$$

where I is the flux density; f_j is the efficiency of radiation; P_j^t is the specific flux of corresponding radiation on quartz at the end of the roll (time t); C_E^t and C_E^o are the concentrations of E_1-centres in quartz at the end of the roll and its head value respectively.

For a particular deposit the dislocation rate of the roll was 3×10^{-6} m/year.

The migration of uranium M (removal or addition) is found from the relationship between the concentrations of uranium determined by the direct method (C_0) and estimated from the distribution of E_1-centres (C_{calc}) [9.4]:

$$M = 100(C_n - C_{calc})/C_{calc} (\%) \tag{9.11}$$

If $M < 0$, then some of the uranium was removed, if, however, $M > 0$, then some uranium was supplied, with $M = 0$ the system was closed. This helps to assess the mineralization in terms of its depth from the distribution of E_1 near the surface.

For instance, the uranium deposit at Blind River in Canada was discovered in the following way [9.3]. The sampling of uranium on the surface gave no encouraging results. On the basis of the distribution of E_1-centres the prospector suspected the presence of rich ores deeper down. When the contract with the company that conducted the survey expired, the prospector drilled wells and found industrial ores.

The age of mollusk shells could be determined by establishing the radiation centres by paramagnetic resonance [9.5]. The age is estimated from

$$T = \frac{D_{acc}}{P_{ex}(\beta + \gamma) + P_{in}(K_\alpha + \beta) + P_c} \tag{9.12}$$

where P_{ex} is the external annual dose of irradiation;

P_{in} is the internal annual dose of irradiation caused by uranium;

P_c is cosmic component;

K_α is efficiency of α-radiation;

D_{acc} is accumulated dose.

9.5 The Thermoluminescence Method

When minerals are heated capture centres in them recombine with the emission of energy within the visible part of the spectrum. This phenomenon is called thermoluminescence. Since the concentration of radiation centres is proportional to the radiation dose absorbed by the mineral during its existence, the intensity of thermoluminescence will also be proportional to the accumulated dose. The amount of luminescence energy released during the heating of the sample is called the light-sum of thermoluminescence. When the light-sum is proportional to dose D of the ionizing radiation, the exposure time of irradiation or the age of object t can be determined from the rate of the accumulation of the dose (the dose rate) P:

$$t = \frac{D}{P} \tag{9.13}$$

Various optically transparent minerals capable of accumulating the light-sum in proportion to the dose can serve as thermoluminescence paleodosimeters. Natural quartz is most widely employed as paleodosimeter, since it has some advantages over other minerals, viz. high occurrence, transparency, adequate dosimetrical properties, and chemical and physical stability. A common quartz paleodosimeter, which is used to date Quaternary deposits is a mineral fraction of quartz with the same size of grains. For the calculation the average light-sum $\overline{S}$ is used. It is established by means of measuring the thermoluminescence S_i for n grains:

$$\overline{S} = \sum_{i=1}^{i=n} S_i$$

An important characteristic of a thermoluminescence paleodosimeter is the saturation light-sum S_∞, which is the limiting light-sum that cannot be exceeded for a given dose rate. For quartz S_∞ is reached at a dose 10^6 rad, which corresponds to the irradiation period of several million years from the natural sources with the abundance concentration of radionuclides [9.5, 9.6, 9.7]. Other authors give this value as substantially lower and corresponding to less than 400 thousand years.

The prerequisites of thermoluminescence dating of Quaternary deposits from quartz grains can be reduced to the following three items.

1. During the weathering of rocks and transportation of fragments on the earth's surface when irradiated by solar ultraviolet light the accumulated light-sum in the minerals is discharged.

2. After the burial of fragmental material and its isolation from the ultraviolet radiation of the Sun a new light-sum begins to accumulate in the material grains due to the natural radionuclides contained in the mineral and cosmic radiation.

3. The ratio of the absorbed radiation dose recorded by the light-sum to the dose rate produces the exposure time for the minerals on the day surface or the age of the dated horizon. The effective dose rate P_{eff} is [9.5]

$$P_{eff} = P_\gamma + K_1 P_\beta + K_1 K_2 K_3 P_\alpha + P_{backgr} \tag{9.14}$$

where P_γ, P_β and P_α are the effective dose rates from γ, β, and α-radiation of the rock enclosing the grains of quartz, K_1 is the relative content of quartz in the rock, K_2 is the relative contribution of the α-saturated layer to the paleodose, K_3 is the efficiency of light-sum conservation from alpha radiation, P_{backgr} is the effective dose rate from the cosmic background.

The formula takes into account the contribution of the type of radiation in producing the efficient dose rate. Therefore, the value of P calculated from the concentration of natural radionuclides proved to be several times greater than P_{eff}.

To determine P_{eff} in practice gamma radiation is measured with field radiometers. To measure thermoluminescence requires a kiln to heat the sample and a photomultiplier to record the intensity of glow. The recording is done with a recorder to control the temperature and light-sum. The paleodosimeter is calibrated after the annealing process by consequent irradiation from a γ-source (e.g. ^{137}Cs with E_γ = 0.661 MeV; or ^{60}Co with E_γ = 1.1 MeV) with doses of different value. After each irradiation the light-sum in the region of the temperature maximum of the thermoluminescence curve is read. For quartz this maximum is around 300 °C. The paleodose is obtained from the calibration curve.

As mentioned above, the calibration does not adequately simulate natural radiation conditions and is the drawback of the thermoluminescent dating method.

There is one method which can dispense with calibration by artificial radiation [9.7]. The age t is determined by the formula:

$$t = \frac{1}{kP} \ln \frac{S_\infty - S_0}{S_\infty - S} \tag{9.15}$$

where S_∞ is the saturation level determined experimentally from old samples; S_0 is the 'zero-point' established by irradiating the sample with an ultraviolet lamp; P is the dose rate at the sampling point found with a field scintillation gamma radiometer; k is the radiation sensitivity, which is established for samples of quartz with a known age by applying the calculation method.

The method requires improvement.

9.6 The Track Method

Spontaneous fission is typical of heavy nuclei at the end of the periodic table and is characterized by its own half-life T_s for each radionuclide [9.8].

In nature thorium and uranium isotopes undergo spontaneous fission. The fission of heavy nuclei induced by cosmic rays and neutrons is insignificant in comparison to spontaneous fission.

The rate of spontaneous fission is several orders lower than that of alpha decay for the same isotopes of uranium and thorium. The most common radionuclides with large T_s is ^{238}U:

	^{238}U	^{235}U	^{232}Th
λ_α, year^{-1}	0.1551×10^{-9}	0.9859×10^{-9}	0.0494475×10^{-9}
λ_s, year^{-1}	8×10^{-17}	3.8×10^{-18}	10^{-22}

Since the rate of spontaneous fission of the most common isotope ^{232}Th is infinitesimal, most of the spontaneous fissions of the nuclei observed in rocks is due to the fission of ^{238}U. Fission results in large number of fragments, e.g. isotopes of krypton and xenon, ^{90}Sr, ^{87}Rb, ^{129}I, ^{99}Tc. Their kinetic energy varies from 30 to 105 MeV, which is sufficient for the formation of tracks in the mineral structures.

Primary tracks can only be observed under an electron microscope. In order to see them with an optical microscope the mineral is etched. Moreover, the composition of the etching reagent is selected so that only the zones around the primary tracks are dissolved without affecting the remainder of the mineral. Knowing the concentration of uranium in the mineral grain and after estimating the density of spontaneous fission tracks, it becomes possible to determine the age of the mineral, or, to be more exact, the time of the latest thermal metamorphism, since the tracks, when heated, undergo annealing. The age is then calculated as

$$t = \frac{2.3}{\lambda_\alpha} \log\left(\frac{N\lambda_\alpha}{U\lambda_s} + 1\right) \tag{9.16}$$

where λ_α and λ_s are the rate constants for α-decay and spontaneous fission respectively, N is the number of acts of spontaneous fission per unit volume of mineral (number of tracks); U is the concentration of uranium.

Spontaneous fission tracks can be distinguished from other non-nuclear crystal defects. They are straight, restricted in length, have a random orientation in the mineral, and are subject to annealing.

The concentration of uranium can easily be determined by tracks of the induced fission of uranium. When the sample has been annealed, it is irradiated by thermal neutrons and the tracks of induced fission of ^{235}U are estimated. Owing to the constancy of the ratio $^{238}U/^{235}U$, the concentration of ^{238}U is calculated from ^{235}U. This method can be applied to establish the age using micas no older than 300-350 mln years, zircons up to 400 mln years old, monazites no older than 300 mln years, and apatites and garnet about 100-200 mln years old. In more ancient minerals the radiation effects can disappear due to weak metamorphosis and superimposed processes that are not manifested in mineralogical terms [9.8, 9.9-9.11].

The advantage of the method is its simplicity and the ability to analyze extremely small amounts of the mineral (fractions of milligrammes). The lower limit of the method depends on the material. The best results were obtained with natural acidic glasses. For instance, obsidians are useful to date ages from tens of thousands to tens of millions of years.

The annealing of tracks in the course of the geological history of a mineral, though restricting the age limit, makes it possible to reconstruct the thermal history of the mineral. The method of tracks may be combined with the argon method. For instance, the age of micas from the pegmatites of Northern Karelia was established by the tracks of spontaneous fission to be approximately 700 mln years, while using potassium-argon dating it was 1760 mln years. The discrepancy reveals that the mineral age was 1760 mln years, but about 700 mln years ago the rock was heated enough to eliminate the earlier tracks [9.11]. It can also be said that the temperature to which the mineral was heated was less than 900 °C otherwise the argon would not have been preserved. Nor could the temperature have been below 400 °C (as tracks would have been left) [9.12].

Fission tracks are also used to determine the forms in which the uranium exists in a mineral, and the direction of its migration. The concentration is established from the tracks of the induced fission with the method given above. The tracks of spontaneous fission yield the distribution of uranium in the remote past (no earlier than 100 thousand years). The induced fission tracks show the recent distribution of uranium. Differences between the patterns describe the migration of uranium at comparatively recent times, and enable us to discover the character and direction of the migration.

The form of the cross-section of the primary tracks depends on the structure of the mineral and makes it possible to identify the structural characteristics in individual small crystals.

9.7 The Pleochroic Halos

Pleochroic halos are spherical dark-coloured zones in minerals which form around inclusions of radioactive species. In polarized light they have poleochroic properties, hence their name, and are frequently found in biotite, fluorite, amphibole, albite, tourmaline, chlorite, cordierite, and other minerals. Halos originate around the microinclusions of radioactive minerals: uraninite, pitchblende, thorianite, thorite, monazite, orthite, cyrtolite, and others. The emergence of halos is due to radiation changes in minerals induced by α-particles of radionuclides, which are in microinclusions. In halos around inclusions of strongly radioactive minerals (e.g. uranite) we can sometimes find jointing or isotropism. The reasons for the colour of pleochroic halos have not yet been clarified. It is thought that the colour is associated with the ionization and excitation of the molecules and the emergence of coloured centres.

The character of the chemical reactions in these centres has also not been clarified. Presumably, the colour is due to the transition of iron from its ferrous to its ferric varieties, the decomposition of water, the formation of neutral metal atoms, etc. Heating minerals also results in the loss of colour by the halo (annealing).

The size of the halos radii ranges between 15 and 50 μm and corresponds to the trace of α-particles in minerals. We mentioned in Chapter 1 that the length of an α-particle path (R) grows with the rate of nuclear decay (λ), $\ln R = A + B \ln \lambda$ (Geiger-Nuttall's law). Therefore halos with large radii yield the most short-lived decay products of uranium and thorium: ^{212}Po and ^{214}Po. It is usual for a halo to consist of several rings, which are separated by dark lines. Each ring is correlated with α-particles that have a particular path length. The outer rings correspond either to ^{214}Po in purely uranium minerals, or to ^{214}Po if thorium is present. The difference in the path lengths of ^{212}Po and ^{214}Po in minerals is 8-10 μm, which can be easily detected and thus indicate the presence of thorium. In some biotites halos, radii of 8.6 and 5.2 μm were found. These could not have belong to the decay products of uranium and thorium and are attributed to α-radiation of ^{146}Sm and ^{156}Gd. The darker outer part of the ring is accounted for by the maximum ionization effect that occurs at the end of an α-particle path.

The intensity of the halo darkening is due to two factors, viz. the α-activity of the inclusion and the duration of irradiation. For instance, estimates indicate that a distinct ring can appear from an inclusion with 10^{-17} g of radium within 300 mln years. This amount of radium is actually present in microscopic inclusions of zirconium, monazite, apatite, to say nothing of uranium minerals proper.

Over time the density of the colour in the halo increases to a certain limit (saturation level), after which it remains constant. Further irradiation can invert or bleach the colour.

Pleochroic halos are subject to annealing, i.e. they disappear with heating. Thus, the pleochroic halos that we see in nature appeared since the rock was last heated. Radioactive irradiation of a mineral leads to the material's loosing and isotropy. Hence, halo zones are susceptible to the circulation of porous solutions. Thus, a study of pleochroic halos in Precambrian hydrothermal-metasomatic uranium ores revealed that halos are replaced by secondary minerals (chlorite, stilpnomelane, hydromicas, and iron hydroxides [9.13]). The diffusion limits with these transformations are insignificant, which led to the preservation of distinct external boundaries of the halos, and often, to the inheritance of the zonal structure of the halos. The processes taking place within the halos are perhaps the reflection of those associated with the formation of ores.

9.8 The Metamict Changes in Minerals

The metamict state is typical of many minerals. In most cases it is manifested in complex oxides, containing tantalum, niobium, titanium, and rare-earths (brannerite, euxenite, fergusonite, polycrase, pyrochlore, samarskite, etc.). Metamict properties are also shared by zircons, thorites, orthites, and thorogummites.

The metamict state occurs when a mineral either partially or wholly loses its crystalline structure and becomes amorphous (isotropic).

All metamict minerals are to a certain extent radioactive. At the same time many highly radioactive minerals do not show any metamict properties (e.g. autunite, carnotite, monazite, thorianite, uraninite). The decisive factor for metamict decay seems to be the radioactive radiation of uranium, thorium, and their decay products, mainly α-particle and recoil nucleus [9.13].

Apparently, the mineral partially melts in the zones adjoining the tracks of α-particles and recoil atoms. Depending on the composition of mineral, solidification results in the formation of an amorphous mixture of oxides (e.g. SiO_2 and ZrO_2 in zircons), or a glass of the primarily compound (for instance, pyrochlore). When solidification yields a crystalline substance, no metamict characteristics are found in spite of the high content of the radioactive elements in the mineral (uraninites). When heated a metamict mineral often converts into a crystalline phase, the recrystallization temperature depending on both the content of the mineral, and its metamict character (e.g. 300-350 °C for pyrochlores, and 700-1300 °C for zircons).

Since the metamict nature of a mineral correlates with its α-activity and the time that has passed since the latest annealing, attempts have been made to date the thermometamorphism of the rocks from the metamict properties of zircons and their α-activity. The metamict nature was determined from the parameters of the elementary cell (by the decrease in X-ray diffraction). The data did not always readily lend themselves to interpretation. One of the reasons for lowering the radiation age is the heterogeneous distribution of uranium and thorium in the mineral. The concentration of uranium and thorium in the metamict phase is always several times greater than in the crystalline one, for which the parameters of the elementary cell are established, whereas the activity A and dose D are totalled for both semimetamict and metamict zircons. Isomorphic impurities can also distort the results [9.14].

9.9 The Role of Radiolysis in Geochemistry

In radioactive and water saturated parts of the earth's crust conditions for the natural radiolysis are created. Such conditions can emerge within uranium or thorium mineralization, in radium-bearing brines at interfaces with oil or gas in the deposits of potassium salts, etc. (see Ch. 3). The study of natural radiolysis is of interest for the interpretation of the geochemical situation where there is high natural radioactivity. A knowledge of these processes is essential for forecasts of the geochemical consequences of burying highly active waste from the nuclear industry in geologic formations (Ch. 8).

The geochemical effects related to the radiolysis of underground water are most apparent in zones of hindered water exchange [9.14]. We often find here unusually high concentrations of oxygen, sulphates and dissolved uranium, which do not correspond to the physical chemistry of the medium. A direct correlation between the components and the content of metals, and the radioactivity of the medium has been established [9.15]. These facts cannot readily be interpreted as due to other processes than radiolysis.

Natural radiolysis becomes noticeable when doses of ionizing radiation absorbed by water are sufficient for its radiolytic decomposition. The maximal yield of the decay products of water decomposition G as a result of ionization and excitation is about 12 molecules/100 eV. With the dose of 10^6 J/kg approximately 1% of the original amount of water is decomposed. When the value G = 12 molecules/100 eV, some 2.4% is decomposed. In Table 9.1 the absorbed doses for the principal types of rocks of the earth's crust of different age are given. The data in the table show that even for the clark content of radionuclides in rocks for the period of time of 100 mln years the absorbed dose is sufficient for a noticeable sequence of radiolytic processes. For

Table 9.1 Mean Absorbed Doses for the Main Types of Rock in the Earth's Crust [9.13]

Item	Sedimentary rocks			Igneous rocks	
	clay	sandstone	carbonaceous	granites	basalts
Contents of elements (according to K. Turekyan and K. Vedepol), %					
U	3.7×10^{-4}	4.5×10^{-5}	2.2×10^{-4}	3×10^{-4}	1×10^{-4}
Th	1.2×10^{-3}	1.7×10^{-4}	1.7×10^{-4}	1.3×10^{-3}	4×10^{-4}
K	2.66	1.07	0.27	3.36	0.83
Rock parameters:					
n, %	21	10	24	1.7	1.7
d, g/cm^3	2.5	2.5	2.2	2.7	2.9
The absorbed dose (in J/kg) over a period of time, years:					
1	2.2×10^{-2}	0.4×10^{-2}	0.8×10^{-2}	2.4×10^{-2}	0.7×10^{-2}
10	2.2×10^{-1}	0.4×10^{-1}	0.8×10^{-1}	2.4×10^{-1}	0.7×10^{-1}
10^4	2.2×10^7	0.4×10^7	0.8×10^7	2.4×10^7	0.7×10^7
10^8	2.2×10^6	0.4×10^6	0.8×10^6	2.4×10^6	0.7×10^6
10^9	2.3×10^7	0.4×10^7	0.8×10^7	2.4×10^7	0.7×10^7
4×10^9	11.1×10^7	1.8×10^7	4.6×10^7	12.3×10^7	3.6×10^7
5×10^9	15.3×10^7	2.4×10^7	6.5×10^7	15.9×10^7	5.0×10^7

relatively high concentrations of radionuclides radiolytic geochemical effects will occur within shorter time spans.

The products of water radiolysis, will oxidize elements in lower oxidation states or reduce those in the highest oxidation states. As a result the mobility of the elements will change and their exchange will increase in the rock-water system.

An estimate of the total amount Q_i of i-th product of radiolytic transformation is expressed by (9.17) and (9.18) [9.15]:

$$Q_i = k_n \frac{G(i)D'M}{N_A} \text{ g/g} \tag{9.17}$$

where D' is the absorbed radiation dose in underground water, eV/g; M is the molecular (atomic) mass; N_A is Avogadro's number; k_n is the coefficient of the transition from J/kg to eV/g (1 J/kg = 6.24×10^{15} eV/g)

$$D' = \frac{n\rho}{\delta} Rt\, c \tag{9.18}$$

where c is the concentration of radionuclide in water; t is time, years; n is porosity, %; ρ is the density of water-enclosing rocks, g/cm³; δ is the water-saturation of rocks, %; R is the dose rate in rocks J/kg.

In radium-bearing underground water the radiolytic yield of oxygen for the Middle Carboniferous (C_2) water-bearing complex of Dnieper-Donetsk depression is

$$D' = \frac{n\rho}{\delta} Rt\, c = 0.048 \times 3.73 \times 10^{25} \times 3.2 \times 10^{8} c = 5.73 \times 10^{32} c,\ \text{eV} / 1$$

$$Q^{C_2}(O_2) = \frac{0.3 \times 5.73 \times 10^{32} \times 32c}{10^2 \times 6.02 \times 10^{23}} = 9.1 \times 10^{7} c,\ \text{g} / 1$$

For the Upper Carboniferous (C_3) water-bearing complex $Q_1^{C_3}(O_2)$ = $1.2 \times 10^8 c$ g/l. With $c = 1 \times 10^{-9}$ g/l ^{226}Ra $Q_1^{C_2}(O_2) = 9.1 \times 10^{-2}$ g/l; $Q_1^{C_3}(O_2) = 1.2 \times 10^{-1}$ g/l = 120 mg/l.

The actual content is almost one order lower than what was calculated, which is due to the consumption of oxygen formed during the radiolytic decomposition of water in various oxidizing processes.

The geochemical validation of the calculations comes from the actual content of oxygen in underground water where there is restricted water-exchange. For instance, water in the oil and gas deposits of the Dnieper-Donetsk depression contains molecular oxygen in 90 out of 270 drilled wells. Its content in free gases varies from 0, n to 20% of volume, while in some cases it was as high as 50% (up to 50-300 cm³/l) [9.14]. Figure 9.1 is an illustration to this phenomenon.

The amount of hydrogen yielded by the radiolysis of underground water in clayey carbon rocks with $G(-H_2O)$ = 5 molecules/100 eV and G(H) = 2.5 molecules/100 eV, is calculated to be 600 thousand tonnes per km³ of rock. The contemporary content of hydrogen in these rocks is much lower because of the numerous reactions due to the hydrogenation of organic compounds, binding to nitrogen, reduction of oxides, bacterial utilization, etc.

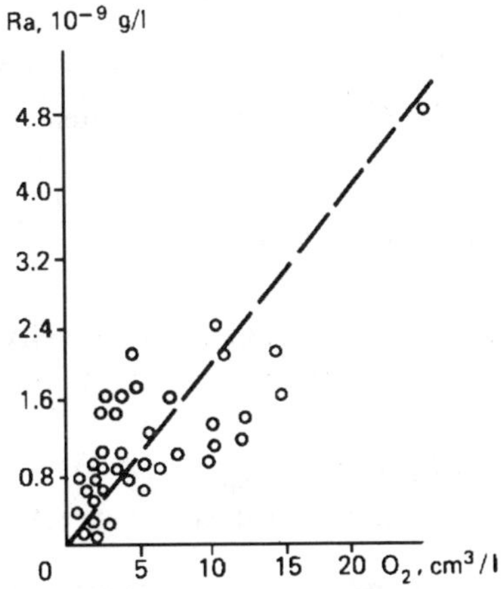

Fig. 9.1. Interrelation between the content of radium and oxygen in the underground waters of the Dnieper-Donetsk trough (according to L.N. Gutsalo, 1971)

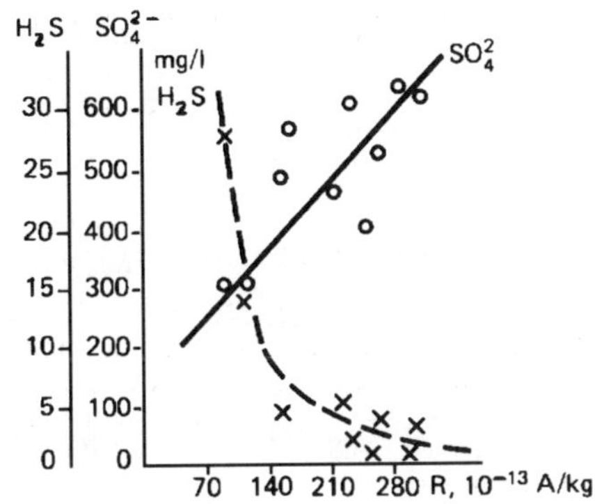

Fig. 9.2. Dependence of SO_4^{2-} and H_2S content on radioactivity in the underground waters of uranium ore manifestation (according to I.E. Vovk, 1974)

An appropriate example of the accumulation of radiolytic hydrogen is found in deposits of potassic salts. In a number of deposits (Strasfurt, Solikamsk, Berezniki, etc.) the content of hydrogen in freely evolving gas in mining was as high as 30-50% and was usually associated with carnallite beds.

The radiolytic decomposition of organic matter dissolved in water transforms it almost completely into carbon dioxide. The radiolysis products react with dispersed organic matter in water-enclosing rocks to induce redox reactions of hydrogenation and dehydrogenation, carboxylation and decarboxylation, and so on. Moreover, the main decomposition products will include methane and carbon dioxide. Various hydrocarbons, ammonium and gaseous nitrogen, hydrogen sulphide, carboxyl acids, carbonyl compounds, phenols, etc. can form.

The data available today prove that sulphides can oxidize in the earth's interior where oxidizing conditions are absent, and can hardly be interpreted in terms of processes other than radiolysis. Radiolytic sulphates are readily formed in underground waters of uranium deposits with sulphide mineralization (Fig. 9.2). The radiolysis of underground waters also provides a satisfactory explanation for the emergence of sulphates in radium-bearing oil waters. These processes result in a separation of a large number of minor elements in the rocks and promote the formation of ore-bearing brines.

The experimental study of radiolysis in the irradiation of natural heterogeneous systems containing water has proved that there is a substantial change in the physical chemistry and the removal of elements from rocks into water. Thus, the irradiation of sediments of the Black Sea increases the content of minor elements in the pore solution compared with the control samples that were not irradiated: I, 16-30 times, Cu, 2-10 times, Pb, 3.47 times, and Zn, 5-12 times. The content of CO_2 and H_2 became very high [9.14].

REFERENCES

9.1. Nesmeyanov, A.N., 1978. Radiochemistry (2nd ed.). Khimiya, Moscow.
9.2. Marfunin, A.S., 1983. Radiation Centres in Minerals.—Izv. Acad. Nauk SSSR, No 5, 100-112.
9.3. Marfunin, A.S., 1975. Spectroscopy, Luminescence and Radiation Centres in Minerals. Nedra, Moscow.
9.4. Moiseev, B.M., 1987. Natural Radiation Processes in Minerals, Nedra, Moscow.
9.5. Isotopic and Geochemical Investigations in Baltic Regions and Belorussia, 1986. Tallinn.
9.6. Vlasov, V.K., Kulikov, O.A., 1977. Calculation of Dosed Velocity Accumulation at Thermoluminescence Dating.—Izv. Acad. Nauk SSSR, **233**, No 3, 467-470.
9.7. Hute, G.I., Punte, P.M., Smirnov, A.V., 1977. Methods of Thermoluminescent Dating in Geology.—Izv. Acad. Nauk SSSR, ser. geol., **4**, issue 26, 281.
9.8. Shukolyukov, Yu.A., 1970. The Fission of Uranium Nuclei in Nature. Atomizdat, Moscow.
9.9. Timothy M. Lutz and Gomaa Omar, 1991. An Inverse Method of Modeling Apatite Fission-track Data.—Earth and Planet. Sci. Lett., **104**, No 2/4, 181-195.
9.10. The Track Method in Geology and Geography, 1987. USSR. Acad. Sc., Vladivostok.
9.11. Komarov, A.N., 1968. Radiographic Methods in Mineralogical and Geochemical Investigations.—Izv. Acad. Nauk SSSR, **1**.
9.12. Karpenko, V.S., 1963. Radioactive Halos Near Uraninite Inclusions.—The Questions of Applied Radiogeology. Gosatomizdat, Moscow, 157-173.
9.13. Makeev, A.F., 1981. Radiation and Chemical Transformations of Zircons and Their Use in Geochronology. Nauka, Leningrad.
9.14. Vovk, I.F., 1979. Radiolysis of Underground Waters and Its Geochemical Role. Nedra, Moscow.
9.15. Gutsalo, L.K., 1971. Water Radiolysis as a Source of Free Oxygen in Hydrosphere.—Geochemistry, No 12, 1473-1481.

10 Field Radiometric Techniques

'What should be of primary concern in Russia is the research into that part of the world's reserves of radioactive substances which are available in Russia.'

V.I. Vernadsky

10.1 Field Radiometric Methods

All natural radionuclides with the exception of ^{238}U, ^{232}Th and some isotopes are found at extremely low concentrations (Ch. 3). The most sensitive methods of analysis are the radiometric, viz. measuring the radioactivity. In the laboratory the radioactivity is measured without the preliminary chemical treatment of samples (radiometric methods proper), as well as with the chemical separation of individual radionuclides (radiochemical methods). All methods of measuring radionuclides in the field are exclusively radiometric.

Most field radiometric methods are based on measuring gamma radiation which is the most penetrating radiation. Beta radiation is sometimes measured to localize the pollution of the soil surface by artificial radionuclides. Alpha radiation is measured only in the field emanation method of analyzing radon isotopes.

The Field Gamma Method

The sensitivity of the method depends on the activity of the radiating objects, the thickness and composition of the inactive layers covering them, as well as the sensitivity of the instruments. Calculations show that almost 99% of the gamma photons emitted by an infinite radioactive bed are absorbed by an inactive layer with a density of 2-25 g/cm^3 and a thickness of the order of 50 cm. Practically all the gamma radiation from the Earth's surface is absorbed in the air at a height of 600-700 m when measuring with standard exploration equipment.

When measuring gamma radiation in the field, the radiation coming from rocks and soils as well as the background radiation, due to cosmic radiation and from the instruments themselves, will affect the results. There are various ways of estimating the background, e.g. measuring over the surface of a water reservoir, measuring the change in the intensity of gamma radiation above the same place, and so on.

In the field two types of measuring gamma radiation are used, viz. the integral and spectrometric methods. The first is to measure the overall gamma radiation. This method cannot be used to assess the contribution of individual radionuclides to the total flux of gamma photons. Integral measurements are expressed in units of dose rate, as a rule, in micro-roentgen/hour (μR/h) [10.1-10.3].

The gamma spectrometric method enables us to determine the activity of separate gamma radionuclides from the energy spectrum of the radiation. The energy spectrum is the distribution of the number of gamma photons as a function of radiation energy. Gamma spectrum measured in natural conditions is fairly complex. They are the results of the superimposing the discrete spectra of the individual radionuclides present in the Earth, and gamma photons dispersed by the rocks. The most probable physical absorption of natural gamma radiation by rocks is reduced to a multiple Compton scattering of photons up to the energy of 0.2-0.05 MeV and their subsequent absorption in the photoelectric effect. The scattering of gamma radiation changes the primary energy spectra of the radionuclides, upon which a continuous spectrum of scattered gamma photons is superimposed. At low energies scattered radiation is predominant. This is why primary gamma radiation with energies less than 0.2 MeV cannot be distinguished against the background of scattered radiation even at the Earth's surface. Further from the Earth surface, the number of spectral lines decreases and shifts to the high-energy range of the spectrum. In the field gamma measurements of natural radionuclides only those parts of the spectra are used where there are distinct peaks with high gamma radiation energies making it possible to characterize individual decay series. Thus, to determine ^{232}Th, one of its decay products is used, ^{208}Tl(E_γ = 2.62 MeV); while to establish ^{238}U or ^{226}Ra we use ^{214}Bi(E_γ = 1.76 MeV). It is common in the field to use 4-channel spectrometers in which one of the channels is set to record the integral gamma radiation, while the other three are adjusted to the intervals of energy corresponding to ^{208}Tl(2.62 MeV), ^{214}Bi(1.76 MeV) and ^{40}K(1.46 MeV).

Disequilibrium in the decay series can result in a discrepancy between the measurements and the true concentrations of uranium and thorium. In the laboratory, discrepancies of this kind for gamma spectrometry are quite natural. The probability of an error due to disequilibrium in decay series when measurements are taken in the field is somewhat levelled out by the great volumes of rocks acting upon the detector during the measurement. Thus, even in the case of point measurements on the Earth's surface the mass of rocks affecting the detector is estimated as hundreds of kilograms. Further from the Earth's surface and depending on the distance h between the radiator and detector, the radius R of the hemisphere of the rocks acting upon the detector will change [10.3] thus,

h, m	10	30	50	70	100	150
R, m	60	120	170	215	280	360

Thus, the further away the detector is from the surface, the larger the average mean value. For instance, soil and rock horizons at the surface may have opposite deviations from the radioactive equilibrium due to the removal of individual radionuclides from one horizon and the accumulation in others. If this occurs within a few tens of centimeters, the horizons will send the same signal to the detector corresponding to radioactive equilibrium. The most reliable measurements will be for thorium due to the relatively short migration paths of some of the members of its decay series. Measurements of uranium may have substantial errors because of disturbances in the equilibrium between uranium and radium, and the migration of uranium beyond the volume of rock being studied.

The way the detector is moved defines the mode of the field method, viz., survey on foot, carborne, airborne, gamma log. The survey on foot is the simplest and easiest. It can be used in any locality and often accompanies a geological survey. The operator also can carry out an accurate geological conjunction of measurement points on the terrain, and, when necessary, take specimens for a more detailed analysis in the laboratory. Portable instruments should not be too heavy, should function reliably in high humidities, and should operate within a wide range of temperatures. Prospecting also requires the extensive use of instruments that only operate in integral conditions. When more elaborate research is conducted, portable spectrometers are used.

The airborne mode involves the continuous recording of gamma radiation aboard a plane or helicopter. The method is more productive than the other methods, and can be used to investigate gamma activity over extensive areas at high speed, as well as to survey places where the other methods cannot be applied. Since the measurements are conducted at a considerable distance from the objects under investigation, average concentrations of elements for large masses of rocks (10^4-10^6 tons) are found on the one hand, while on the other, there does exist the risk of missing smaller objects. The layer of air between the detector and the Earth's surface attenuates the gamma radiation and restricts the altitude of the flight. The minimum height is set by the rules governing flight safety. In many cases a height within the limits of 25-75 m is used. The greatest altitude is used in mountain districts, while in the plains the height is the least. A principal requirement in aerial measurements is to maintain a particular altitude of the flight. The measurement of gamma activity in motion and at a considerable distance from the radiator demands special instruments that must be very sensitive and have quite a low inertia. Gamma activity is continuously recorded and automatically processed, and after certain corrections, the results are derived in the form of graphs or maps showing the distribution of gamma activity, the content of uranium, thorium, and potassium.

The carborne mode has much in common with the aerial survey. The measurements are also conducted using a 4-channel gamma-spectrometer with continuous recording. This type of gamma-survey is characterized by a greater sensitivity, but lesser productivity than aerial surveys, though it is restricted if transport cannot reach the site. It is usually employed with foot surveys, when more detailed investigations outside the itinerary should be used.

The borehole gamma logging or the research in the gamma activity in boreholes has its own features. Measurements are taken when the detector is surrounded with radiating rocks. The affect of cosmic background is minimal. The radiation detector is inserted into an air-tight drilling contrivance, which is moved along the borehole with the help of a flexible cable. The registration is mainly done in the course of a continuous recording.

The Field Radon Method (Emanometry)

The method measures radon isotopes in the soil air. The measurement is carried out by a special instrument called an emanometer, which consists of an air-tight emanation chamber into which soil air containing radon is introduced, and the measuring instrument. The chamber functions as a detector of alpha radiation at the same time. Besides, the detector includes a portable pump, with the help of which soil air is pumped out of the blastholes, sampler, and a drier. The blastholes are usually pierced into the depth of 0.8-1.0 m. After a sampler has descended into the blasthole, measurements are taken to avoid the penetration of atmospheric air. Radon isotopes are the only radioactive gases within the hypergenesis zone. They are all alpha radiators, and hence the whole of measured alpha activity will be due to radon. The background is only determined by the contamination of the chamber.

10.2 Field Radiometric Methods for Geological Mapping

Thorium, uranium, and potassium can easily be discovered in rocks by applying radiometric methods even for their abundance concentration. This means that they can be used as natural tracers, by utilizing cheap and express radioactive methods. The main condition for the use of radiometric methods in the field is a significant difference in the radioactivity of objects that are compared: ore or enclosing rock, two different kinds of rocks, etc.

Field radiometric methods are employed in the exploration of radioactive mineral resources, as well as for geological mapping. The decisive factor which should always be taken into consideration is the statistically significant difference of gamma activity, or the concentration of uranium, thorium, potassium, and radon in geological objects to be compared. A detailed description of the statistical methods of processing the data can be found in corresponding guides.

For a successful use of radiometric methods in geological mapping it is necessary to have all the data on the geochemistry of uranium, thorium, and potassium in the rocks being mapped. The prospecting should be accompanied by a geological survey, for which radiometric methods are subsidiary. Information on the geochemistry of uranium and thorium is given in Chapter 4. Below we present the basic patterns of the distribution of radioactive elements in rocks, which can be used to plan the survey and interpret the radiometric mapping.

1. In the normal igneous rock series of a single series of alkalinity, the content of uranium, thorium, and potassium will increase gradually from basic to acid rocks.

2. An increase in alkalinity (potash content) is accompanied by a growth in the concentrations of uranium and thorium.

3. An increase in the content of radioactive elements from the early to the later phases, as well as to the marginal facies, is characteristic of multiphase differentiated intrusions of granites.

4. Granitic intrusions of different ages can differ in their radioactivity. In most cases young intrusions are more radioactive than older ones of the same composition.

5. Higher radioactivity characteristic of metasomatic alterations of a rock.

6. Hydrothermal changes of rocks are usually accompanied by an increase in the content of uranium or thorium.

7. Tectonic dislocations are more radioactive than enclosing rocks due to hydrothermal changes in rocks in the tectonic zone, and the accumulation of radon in cracks.

8. Among sedimentary rocks, clays are more radioactive than sandy or carbonate rocks.

9. Rocks of the same lithologic type of higher radioactivity may be associated with organic matter, phosphorus, and sometimes iron hydroxide deposits or zones.

10. Eluvial and eluvial-deluvial deposits inherit the radioactivity of the parent rocks. This makes it possible to conduct a radiometric survey where an eluvial-deluvial mantle rock has developed.

Mapping implies a combined application of radiometric methods, i.e. airborne survey, carborne survey (Fig. 10.1), and radon surveying (Fig. 10.2). All these methods need to be combined with other geological and geophysical methods.

The airborne gamma-ray spectrometer survey can map vast outcrop areas of rocks differing in radioactivity, particularly in plain topographies. An example is the mapping in Western Siberia, which made it possible to contour

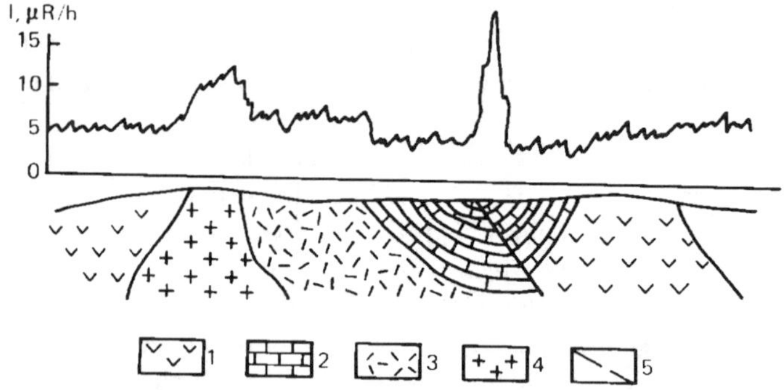

Fig. 10.1. An example of the release of automobile gamma-survey contacts of rocks and tectonic dislocations [10.2]:
1–basic effusives; 2–limestones; 3–acid effusives; 4–granites; 5–tectonic dislocations

intrusions of granitoids with higher radioactivities (15-20 µR/h) than the enclosing rocks (8 µR/h). A similar aerial survey in Eastern Siberia enabled researchers to age the intrusions of granitoids according to their gamma activity, which could not be achieved by other methods. The most ancient Caledonian granites (6-7 µR/h) have the lowest activity, while the youngest Cimmerian ones are associated with the highest activity (10-30 µR/h), and Variscian granites occupy an intermediate position. On the Kolyma plateau the same method made it possible to identify regions characterized by acid effusives whose gamma activity of 8-20 µR/h among basic effusives, whose activity is 2-6 µR/h. To make an airborne survey more accurate in individual localities, and to map mountain regions, survey on foot is more useful. This is true when mapping massifs of igneous rocks, to segregate individual facies, metasomatically altered zones, tracing veins, dykes, and fissure zones. In the Northern Caucasus survey on foot contoured granite intrusions with activities of 20-40 µR/h against the background of the enclosing effusives, whose activity was lower than 15 µR/h under eluvial and deluvial deposits [10.4, 10.5].

An emanometry survey can provide details on geological boundaries hidden by drift beds from 1 to 8 meters thick. This method produces particularly favourable results when mapping fissures with gas accumulations. Good results come from the joint use of radon and gas surveys, especially CO_2 measurements.

Measuring radon in combination with He survey was applied to study contemporary tectonic movements in regions of high seismicity, and can help forecast earthquakes. During the 1966 Tashkent earthquake, Ulomov noticed that the concentration of radon in the underground water wells began to increase several days before the earthquake, rising by 5-7 times immediately before the shock [10.6]. During the shock the radon content in the water dropped

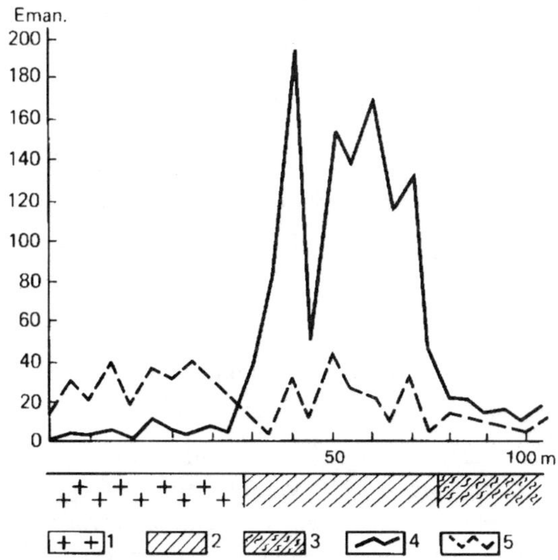

Fig. 10.2. The use of fault zones:
1–granite-gneisses; 2–graphite-carbonate-siliceous slates; 3–micaceous slates; ^{222}Rn concentration; ^{220} Rn concentration

(by approximately 10 times), slowly rising to the original values. It is assumed that long before the shock, fissures began to be exposed in the faults; the pore solutions mixed with the gravitational ones, and radon entered the underground water. At the time of the shock, there was presumably a marked loss of radon. Then its concentration gradually grows to the initial level due to radium decay. There is some reason to believe that in combination with other geochemical and geophysical methods, a radon survey may also forecast volcanic eruptions in regions of active volcanism. Attempts have been made in hydrogeology to apply radon surveys to discover areas of the intrabed and benthonic discharge of underground water [10.7-10.9].

10.3 Field Radiometric Methods in Prospecting for Radioactive Ores

To do this all the geophysical, geological, and geochemical methods are integrated. In all cases radiometric field methods dominate the exploration [10.1-10.3, 10.10]. Irrespective of method the search for radioactive raw materials as well as those of other elements follows the general procedure of: (a) determining the local background; (b) discovering anomalies; (c) selecting ore-bearing anomalies. A.P. Solovov [10.11] expresses the lower abnormal concentration of C_a or I_a in terms of gamma activity as:

$$C_a = \overline{C} + 3S$$

$$I_a = \overline{I} + 3S \quad \text{(in the case of a normal distribution)}$$

$$C_a = \tilde{C} \times \varepsilon^3$$

$$I_a = \tilde{I} \times \varepsilon^3 \quad \text{(in the case of a lognormal distribution)}$$

where $\overline{I}$ and $\overline{C}$ are the mean activities or concentrations; $\tilde{C}$, $\tilde{I}$ are the mean geometrical values; S is the standard deviation; and ε is a standard factor.

A characteristic of anomaly is the excess over the background. An indicator of the contrast range γ, according to Solovov, is the ratio of the amplitude of the anomaly excess over the background to the index of stability, viz.

$$\gamma = \frac{C_{\max} - C_{\text{backgr}}}{S}; \ (C_{\text{backgr}} = \overline{C})$$

or

$$\gamma = \frac{1}{\log \varepsilon} \cdot \log\left(\frac{C_{\max}}{C_{\text{backgr}}}\right); \ (C_{\text{backgr}} = \tilde{C})$$

Airborne Prospecting

Airborne prospecting is the principal initial method in the overall search for uranium. It makes possible in a relatively short period of time to discover those parts of the territory that are promising as far as the organization of exploration on land is concerned.

The most effective use of airborne techniques is in smooth topographies, dry climate, and wide halos of uranium scattering. In humid climates with an abundance of precipitation, poor exposure, and thick forest covering, as well as in mountain regions, the airborne exploration is less effective. A moist layer of friable deposits results in a considerable leaching of uranium, which, in turn, decreases the γ-activity along the surface of dispersion halos. Vegetation also plays a role in diminishing the intensity of γ-field around the anomalous zones. In mountain regions, in addition to the danger of flight, it is very difficult to maintain the same altitude in flights, which leads to distortions in recording and hampers the discovery of ore anomalies.

When looking for uranium ores we often have to deal with anomalies that are irregular in form and have small areas (100-300 m in diameter at the height of the flight). The possibility of finding ore anomalies is also hindered by the nonuniform radioactivity of the anomalies themselves and of the enclosing rocks. This requires the flights to be at the minimal height, so as to single out the anomalous activity against the γ-background of the enclosing rocks.

Before making surface exploration from the air it is necessary to choose routes so as to assess the radioactivity in the rocks in that region and the geomorphological conditions. The explorations are conducted according to parallel itineraries across the extension of the main geological structures. The length of a route is 25-30 km. In mountain regions prospecting occurs with helicopters, while the routes are oriented parallel to mountain ridges. The distance between the routes is determined by the survey scale, so that the distance between them on the map is 1 cm. As a rule operations are conducted with a scale of 1 : 100 000, 1 : 50 000, or 1 : 25 000. Details are made on a larger scale than the usual one (1 : 10 000).

When interpreting γ-anomalies, it is essential to take into account the total of geological, geomorphological, physical, geographical, radiometric, and geophysical data.

When aerogamma anomaly is recognized as promising it becomes necessary to verify it on land.

Carborne Prospecting

Carborne radiometric prospecting is intended to solve the same tasks which were presented as on foot ones. With their help airborne anomalies are verified, as well as the primary assessment of the perspective territory is made. The advantage of airborne surveys over the on foot variety is, as has already been mentioned, in greater productivity and sensitivity of the instruments. There are route and area gamma surveys. The former is mainly employed where reconnaissance is required. Area surveys (scale 1:25 000 to 1:1000) are used in places that are accessible for vehicular transport. They are conducted by means of parallel routes 1-4 km long. The distance between the routes, as in the airborne method, is determined by the scale.

Verification of the carborne anomaly is carried out by the γ-method on foot parallel with the geological and geomorphological survey. When the perspective anomaly is detected, it is studied in greater detail by gamma spectroscopy on foot and lithochemical (uran-metric) methods, opencut mining operations.

On Foot Prospecting

Local and spot surveys are best made on foot. It is a simple and widespread method. This method is employed at all stages in uranium exploration, beginning with reconnaissance and ending with a preliminary evaluation of the deposits, and can be used in those regions that are inaccessible for other radiometric methods.

On foot γ-exploration is most favourable in regions of an arid climate. Friable eluvial or eluvial-deluvial cover can enclose open mechanical and saline dispersion halos and in this connection can also be used for radiometric measurements.

The territories covered by allochthonous deposits are less promising for γ-quests on the surface. In humid climates the upper layer of eluvial-deluvial mass is depleted in radioactive elements as a result of leaching, which accounts for the fact that γ-prospecting on the surface in most cases also proves to be less effective. In these regions borehole γ-prospecting produces better results. The measurements of γ-activity are carried out at the face of a borehole at a depth of 1 metre. In many cases this depth is sufficient to discover dispersion halos below the depleted layer. Aluvial, proluvial and paludal deposits in humid climates can also favour the formation of scattered concentrations of uranium and should be thoroughly studied during the exploration.

An anomaly discovered by on foot γ-explorations is then studied with on foot γ-spectrometry method, borehole surveys, and radon method of prospecting. When the anomaly is really promising, different mining operations are conducted on its territory. They should also be accompanied by the measurement of the γ-activity of rocks exposed by mining.

Depth Gamma Explorations

The surface γ-methods discussed above are characterized by a relatively small depth. Looking for uranium deposits in regions hidden by aluvial drifts not genetically connected with the underlying rocks is practically ineffective by these methods. To increase the efficiency of uranium exploration in closed territories depth methods are employed. The simplest is the borehole γ-survey, which was mentioned in the previous section. It is used to detail surface γ-anomalies and is conducted on a grid. The distance between the holes is determined by the size and character of the anomaly.

To improve the method special self-propelled borehole installations are used which help to obtain blastholes over 20 m deep. Depth explorations are conducted within the perspective areas within the confines of uranium provinces. The principal object of the research are buried dispersion halos.

Operations of this kind can be conducted where there is passable area for motor transport. Drilling is carried out with a network. The distance between the profiles when deep searches of large deposits are conducted were chosen to be 500-700 m, and for small and middle deposits the figures do not exceed 200-300 m, while for some ore bodies it is 100 m and less. The distance between the boreholes along the profile should be less than 50 m. The measurement in shallow blastholes is made by the logging the on foot measurement while in deeper ones it is recorded with automatic instruments and a detector inserted into the drill.

Prior to drilling the depth of the 'representative horizon' is determined. It is the horizon of maximum areal and fairly reliable secondary dispersion halo.

Gamma-Ray Log

In all the stages of the exploration and reconnaissance of uranium deposits accompanied by drilling, the wells are investigated by the γ-logging. Exploration γ-logging consists in measuring the γ-activity of rocks along the depth to discover anomalies and to evaluate the normal values of the γ-field. In better circumstances γ-logging helps to assess more accurately the thickness of the orebody, and the content of uranium in it. Data from γ-logging are used to count the reserves. The true thickness of an orebody (h_{true}) is established according to the formula

$$h_{true} = h_{app} \cdot \sin \alpha,$$

where h_{app} is the thickness determined from the diagram, and α is the angle between the well and the orebody.

The content of uranium is given by:

$$q = \frac{S}{100 k_0 h},$$

where q is the content of uranium, %; S is the area of α-anomaly in the diagram, μR/h; k_0 is a coefficient, μR/h per 0.01% of equilibrium uranium; h is the thickness of orebody, cm.

A coefficient k_0 is determined experimentally for each type of a detector, registering instruments and for the conditions of measurement. To make quantitative interpretation of γ-logging data, a number of corrections are introduced (those concerning the absorption of γ-radiation in borehole solution and pipe casing, humidity, disequilibrium in ores, and the presence of Th).

Radon Method of Prospecting

This method of uranium prospecting is intended for the exploration of radon gaseous dispersion halos formed around uranium orebodies and dispersion halos of uranium and radium.

The emanation method is not widely used today in a detailed exploration and elaboration of anomalies discovered by γ-methods. It proves to be more reliable than surface γ-methods and makes it possible to discover uranium mineralization when the thickness of the drifts not containing mechanical and saline halos, is from 1 to 5 m. If in deposits there is a distinct saline uranium-radium dispersion halo, the depth at which the ore is discovered is increased. Gas halos are formed due to the gas permeability of the drifts covering the mineralization. If the thickness of deposits screening radon (clays, markedly moistened paludal formation, etc.) exceeds 1.5-2 m, the organization of a soil radon survey is ineffective. A propitious for soil radon survey is drier parts of the territory, with the thickness of fairly porous drifts within the limits of 1-5 metres. For outcrops of bedrock, talus, and marsh this method is inapplicable.

During an emanometry survey the activity is measured immediately after the soil air is introduced into the emanation chamber.

In anomalous parts, two-three measurements are taken at each point: directly after radon is introduced, after 2 and 3 minutes, when the whole of Thoron (^{220}Rn) is decayed.

Radiohydrochemical Prospecting

The radiohydrochemical method is intended to discover aqueous dispersion halos, which are formed around uranium ore bodies and their lithochemical halos. It consists in testing underground and surface waters for uranium, its decay products and associated elements. The main interest of uranium decay products for hydrochemical explorations is radon, and also ^{226}Ra and ^{234}U. The set of associated elements depends on the type of deposit. The most usual associates of uranium are molybdenum, lead, arsenic, vanadium, phosphorus, and selenium.

10.4 Radiometric Survey in Prospecting Other Mineral Deposits

The use of radiometric methods to explore prospect deposits of non-radioactive mineral deposits is possible when the radioactive element is genetically or paragenetically associated with the ore elements, or when the orebodies differ from the enclosing rocks by their radioactivity or uranium, thorium, or potassium ratios [10.5, 10.12, 10.13].

The most stable paragenetic associations with thorium and uranium are with the rare-earth elements, yttrium, zirconium, tantalum, and niobium. Associations of uranium and phosphorus, thorium and aluminium, uranium, vanadium, and selenium are also known. Prospecting for such mineral deposits requires radiometric methods at all stages of the operation from the quest for ore-controlling structures and ending with an approximate evaluation of the resources. An effective method is airborne gamma method at reconnaissance and prospecting stage with spot verification of spot survey and borehole logging. The methods used in each case depend on many factors: viz., the scale of bedding, the thickness of covering drifts, the presence of dispersion halos, the concentrations and ratios of the uranium, thorium, and potassium in the orebodies and halos, etc. In all cases gamma spectrometric methods are needed. Practical experience shows that the Th/U, Th/K and U/K ratios provide better contrasting anomalies than total gamma activity. Some examples of the use of field radiometric methods in the exploration of raw materials are given below.

Prospecting for Phosphorites

A high uranium content (10-100 ppm) and low thorium content with respect to the abundance ratio are characteristic of most phosphorite deposits. We can find a positive correlation between the content of uranium and phosphorus, which makes it possible to use radiometric methods to prospect for phosphorites. Contrasting anomalies are found if we use the U/Th ratio. Moreover, radiometric methods can be employed to analyse the P_2O_5 content in raw materials, and to automate H_2SO_4 batching at superphosphate plants [10.10, 10.11].

Prospecting for Bauxite

Bauxites, laterite weathering crusts, have usually high contents of thorium 30-70 ppm, against a background of 5-10 ppm, and a low concentration of potassium (0.5%). The uranium content in bauxite is usually close to the abundance ratio, although occasionally somewhat higher. The search for bauxite calls for the application of gamma spectroscopy. Contouring of anomalies using the Th/K ratio yields the greatest contrast.

Prospecting for Tantalum-Niobium Mineralization

The rare-earth tantalum niobates always contain isomorphic impurities of thorium and uranium. These metal deposits in carbonatites or ancient alkaline metasomatites stand out against the background of enclosing rocks by their higher gamma activity, due to thorium, uranium, and potassium. This enables prospectors employ various modifications of the gamma method to discover and contour the orebodies. Gamma spectrometric measurements yield the structure of the orebodies, the depth of the erosion cross section, and the character of the metasomatic processes in great detail from the Th/U, K/U, and K/Th ratios. Sometimes we can also find a positive correlation between Ta and the radioactive elements in ores, which makes it possible to use gamma spectrometric to evaluate the reserves.

Prospecting for Beryllium

Alkaline metasomatites, which are associated with deposits of Be, usually have higher radioactivity than the enclosing rocks. As a rule, this is due to the higher concentrations of thorium in the pre-ore metasomatites, where the alkalis contain mainly K. In the latest hydrothermal stage the potential of thorium in solutions drops and uranium has more friction. The highest degree of radioactivity of potassium occurs where there is potassium metasomatosis.

Potassium Prospecting

The gamma-activity of potassium-bearing minerals conditioned by the presence of radioactive isotope ^{40}K makes it possible to employ radiometric methods for potassium deposits. Gamma logging can be used here most propitiously. There exists an opinion that gamma logging can yield more accurate results in the calculation of reserves than the chemical testing of core samples.

The Search for Hydrocarbons

Work by many scientists has made it possible to look for oil and gas deposits using airborne gamma methods. The intensity of the gamma field falls over the greater part of an oil and gas deposit. The negative anomaly over the deposit is often surrounded by a ring of slightly higher gamma activity exceeding the background by 10-15%. These facts enable us to use gamma surveys to contour oil and gas deposits. The conditions that are favourable for these activities require research on a wider scale.

REFERENCES

10.1. Novikov, G.F., Kapkov, Yu.N., 1969. Radioactive Methods of Exploration. Nedra, Leningrad.

10.2. Novikov, G.F., 1989. Radiometric Survey. Nedra, Leningrad.

10.3. The Methods of Uranium Deposit Exploration, 1969. Nedra, Moscow.

10.4. Baranov, V.I., Titayeva, N.A., 1973. Radiogeology. Moscow State Univ. Publ., Moscow.

10.5. The Experience of Radioactive Methods Used in Searching and Exploration of Non-radioactive Ores, 1962. Gosgeoltekhizdat, Moscow.

10.6. Ulomov, V.I. Mavashev, B.Z. About the Forerunner of Strong Tectonic Earthquake, 1967.—USSR Acad. Sc. Rep., **176**, No 2, 319–321.

10.7. Basentsyan, M.M., Rudakov, V.P., 1989. Radon, the Forerunner of Spitak Earthquakes (December 7, 1988).—Izv. Acad. Nauk Arm. SSR, Sciences about the Earth, No 3, 64-66.

10.8. Antanasijèvic, R., Milosenic, I., 1990. Correlation Between the Concentration of ^{222}Rn at the Earth's Surface and in Waters with Seismic Activities.—Nucl. Track Rad. Measurem., No 1, 79.

10.9. Hydrochemical Signs of Earthquakes, 1985. Nauka, Moscow.

10.10. The Handbook of Geologist in Prospecting and Exploration of Uranium Deposits, 1989. (Ed. N.P. Laverov). Nedra, Moscow.

10.11. Solovov, A.P., 1987. Geochemical Prospecting for Mineral Deposits. Mir Publishers, Moscow.

10.12. Aerogamma Spectrometric Method of Searching of Ore Deposits (a collection of articles), 1977. Nedra, Leningrad.

10.13. Vorobyev, V.P. et al., 1982. Aerogamma Spectrometry in Geology. Nedra, Leningrad.

Appendix 1

Table of Radioactive Series

Nuclide	Historical name	Period of half-life $T_{1/2}$	Radioactive constant λ, s^{-1}	Character of radiation	Energy, MeV		Free path of α-particles in the air under normal pressure, cm	Maximum free path of β-particles, g/cm^2	Equilibrium mass ratio
					particles	γ-rays			
				Uranium series					
$^{238}_{92}U$	Uranium I	4.51×10^{9} yrs	4.883×10^{-18}	α	4.132(22%)	2.64			
				γ	4.180(78%)	0.048	2.685		1.0
$^{234}_{90}Th$	Uranium X_1	24.10 days	3.32874×10^{-7}	β^-	0.103(34%)	0.043	0.015		
				γ	0.193(66%)	0.047			
						0.0904		0.040	1.44×10^{-11}
$^{234}_{91}Pa$	Uranium X_2	1.175 min	9.8314×10^{3}	β^-	0.600(7%)	0.782	0.200		
				γ	1.500(13%)	0.817		0.67	
				isomeric transition (0.15%)	2.320(80%)	0.396 (isomeric transition)		1.03	4.88×10^{-6}

Appendix 1 (*continued*)

Nuclide	Historical name	Period of half-life, $T_{1/2}$	Radioactive constant λ, s^{-1}	Character of radiation	Energy, MeV		Free path of α-particles in the air under normal pressure, cm	Maximum free path of β-particles, g/cm²	Equilibrium mass ratio
					particles	γ-rays			
$^{234}_{91}Pa$	Uranium	6.7 hrs	2.87×10^{-5}	β^-	0.45(90%)	0.810		0.13	
						0.845			
				γ	1.2(10%)	0.877		0.61	2.51×10^{16}
$^{234}_{92}U$	Uranium II	2.475×10^{5} yrs	8.8743×10^{-14}	α	4.716(26%)	0.050	3.212		
					4.763(74%)	0.117	3.260		5.41×10^{-5}
					4.464		2.958		
					4.482	0.0678	2.976		1.78×10^{-5}
$^{230}_{90}Th$	Thorium	7.52×10^{4} yrs	2.65×10^{-13}	α	4.61(25%)	0.150	3.104		
				γ	4.68(75%)	0.210	3.177		
$^{226}_{88}Ra$	Radium	1602 yrs	1.3814×10^{-11}	α	4.589(5.7%)		0.083		3.36×10^{-7}
				γ		0.188			
					4.777(94.3%)		0.274		
$^{222}_{86}Rn$	Radon	3.825 days	2.0974×10^{-6}	α	5.482		4.736		2.17×10^{-12}
$^{218}_{84}Po$	Radium A	3.05 min	3.788×10^{-3}	α (99.96%) β^-(0.04)	5.996(100%)		4.638		1.18×10^{-15}

$^{218}_{85}At$	Astatine	2.0 s	33.0	α	7.79		7.027		0.136×10^{-18}
					0.350	0.053		0.095	1.02×10^{-14}
					0.670	0.241924		0.240	
$^{214}_{82}Pb$	Radium B	26.8 min	4310×10^{-4}	β^-	0.726	0.320	0.260		
				γ	0.779	0.295217		0.280	
						0.351992			
					5.33	0.0625	3.865		0.75×10^{-14}
				α (0.04%)	5.444(55%)	0.1911	3.993		
$^{214}_{83}Bi$	Radium C	19.7 min	5.864×10^{-1}	β^- (99.96%)	5.505(45%)	0.505	4.062		
						0.60937			
					1.65	0.769			
						3.17(77%)			
						1.22		1.60	
						1.241			
						1.419			
						1.766			1.04×10^{-21}
						0.608			
						0.120			
$^{214}_{84}Po$	Radium C′	1.637×10^{-4} s	4.2343×10^{3}	α	7.6802	1.235	6.869		
						1.761			
						2.090			
$^{210}_{81}Tl$	Radium C″	1.32 min	8.752×10^{-3}	β^-	1.800			0.84	1.97×10^{-19}
						0.0078			4.32×10^{-9}
$^{210}_{82}Pb$	Radium D	21 yrs	9.98×10^{-10}	β^-	0.0167	0.0161		0.0006	
				γ	0.0556(10%)	0.0232		0.005	

Appendix 1 (*continued*)

Nuclide	Historical name	Period of half-life, $T_{1/2}$	Radioactive constant λ, s^{-1}	Character of radiation	Energy, MeV		Free path of α-particles in the air under normal pressure, cm	Maximum free path of β-particles, g/cm^2	Equilibrium mass ratio
					particles	γ-rays			
$^{210}_{83}Bi$	Radium E	5.02 days	1.608×10^{-6}	β^-(100%)	1.17			0.50	2.68×10^{-12}
				$\alpha(5\times10^{-5}\%)$	4.93		3.434		
	Polonium								
$^{210}_{84}Po$	Radium F	138.374 days	5.80002×10^{-8}	α	3.3	0.084		1.67	0.74×10^{-10}
				β^+	4.5(10%)	0.804	2.994		
				γ	5.298		3.83		
$^{206}_{82}Pb$	Radium G	stable	—	—	—	—	—	—	—
				Actino-uranium series					
$^{235}_{92}U$	Actino-	7.13×10^{8} yrs	3.081×10^{-17}	α	4.20(4.2%)	0.094	2.704		7.7×10^{-3}
	uranium			γ	4.393(85.6%)	0.143	2.889		
						0.184			
					4.58(10.2%)	0.289	3.074		
						0.386			
					0.093(45%)			0.0012	
$^{231}_{90}Th$	Uranium Y	25.5 hrs	7.5259×10^{-6}	β^-	0.216(11%)	0.022		0.044	

$^{231}_{90}$Th	Uranium Y			γ	0.302(44%)	0.059	0.080	
	(continued)				0.010			
					0.032	0.063		
					0.0377	0.085		2.98×10^{-1}
					0.0413			
					0.0535		0.107	
					0.0535		0.107	
					0.0578		0.122	
					0.0612		0.167	
					0.0767		0.208	
					0.0805		0.230	
					0.0908			
					0.115			
$^{227}_{89}$Ac	Actinium	21.6 yrs	1.012×10^{-9}	α(1.25%)	4.6(α)(15%)	0.015	3.094	
				β^- (98.75%)	4.95(α)(85%)	0.3	3.455	2.12×10^{-10}
				γ		0.01		
$^{227}_{90}$Th	Radioactinum	18.2 days	4.245×10^{-7}	α	5.672(2.5%)	0.050	4.256	
				γ	5.717(15%)	0.0265	4.309	
					5.742(4%)	0.0306	4.339	
					5.764(20%)	0.0479	4.364	4.98×10^{-13}
					5.815(1%)	0.0550	4.424	
					5.868(2.5%)	0.0606	4.486	
					5.922(1%)	0.0709	4.550	
					5.966(4%)	0.0796	4.602	
					5.988(25%)	0.0848	4.628	
					6.017(4%)	0.0881	4.663	
						0.1004		
						0.1118		

Appendix 1 (*continued*)

Nuclide	Historical name	Period of half-life, $T_{1/2}$	Radioactive constant λ, s^{-1}	Character of radiation	Energy, MeV		Free path of α-particles in the air under normal pressure, cm	Maximum free path of β-particles, g/cm^2	Equilibrium mass ratio
					particles	γ-rays			
$^{227}_{90}Th$	Radioactinum (*continued*)					0.1274			
						0.1322			
						0.1492			
						0.1643			
						0.1729			
						0.2041			
						0.2117			
						0.2194			
						0.2400			
						0.2532			
						0.2734			
						0.281			
						0.304			
						0.322			
						0.353			
						0.377			
						0.426			
						0.466			
						0.513			
						0.578			
						0.638			

$^{233}_{87}Fr$	Actinium K	22 min	5.50×10^{-4}	β^-, γ $\alpha(4.10^{-3}\%)$	1.200	0.095		0.51	4.78×10^{-18}
$^{233}_{88}Ra$	Actinium X	11.4 days	7.163×10^{-7}	α	5.419(3%)	0.0262	3.964		
					5.487(2%)	0.0637	4.041		
				α	5.528(9%)	0.0809	4.088		
					5.526(24%)	0.0991	4.167		
					5.704(53%)	0.1163	4.294		
					5.730(9%)	0.144	4.325		
					5.860(low)	0.1539	4.477		2.94×10^{-13}
						0.1615			
					0.1797				
					0.2318				
					0.2677				
					0.2798				
					0.322				
					0.3476				
					0.444				
				α	6.434(12%)	0.0679	5.178		
					6.559(15%)	0.1234	5.339		
					6.824(69%)	0.1980	5.652		1.17×10^{-18}
$^{219}_{88}Rn$	Actinon	4.0 s	0.1767	α		0.2667			
						0.321			
						0.392			
						0.589			
$^{215}_{84}Po$	Actinium A	1.78×10^{-3} s	3.787×10^{2}	α	7.365(~100%)		6.422		5.36×10^{-22}
				β	$(5.10^{4}\%)$				
$^{215}_{85}At$	Astatine	10^{-4} s	7×10^{3}	α	8.00(~100%)		7.336		1.33×10^{-26}

Appendix 1 (*continued*)

Nuclide	Historical name	Period of half-life, $T_{1/2}$	Radioactive constant λ, s^{-1}	Character of radiation	Energy, MeV		Free path of α-particles in the air under normal pressure, cm	Maximum free path of β-particles, g/cm^2	Equilibrium mass ratio
					particles	γ-rays			
$^{211}_{82}Pb$	Actinium B	36.1 min	3.200×10^{-4}	β	0.571(20%)	0.06524		0.190	
					1.40(80%)	0.0829		0.62	
						0.342			6.22×10^{-6}
						0.4040			
						0.4258			
						0.4871			
						0.7640			
						0.8296			
$^{211}_{83}Bi$	Actinium C	2.15 min	5.348×10^{-3}	α	6.621(82.6%)		5.419		
				(99.68%)					
				(0.32%)	6.274(17.4%)	0.347	4.977		3.72×10^{-17}
$^{211}_{84}Po$	Actinium C′	0.52 s	1.333		6.34(0.07%)	0.540	5.059		
					6.57(0.48%)	0.565	5.353		
					6.90(0.57%)	0.870	5.786		4.78×10^{-22}
					7.434(98.88%)		6.517		
$^{207}_{81}Tl$	Actinium C′′	4.79 min	2.412×10^{-3}	β	0.630(0.5%)	0.8700		220	8.07×10^{-17}
					1.500(99.5%)			0.680	
$^{207}_{82}Pb$	Actinium D	stable							

Thorium series

$^{232}_{90}Th$	Thorium	1.4×10^{10} yrs	1.5813×10^{-18}	α	3.905(20%)		2.436		1.0
				γ	3.98(80%)		2.503		
						0.075			
$^{228}_{88}Ra$	Mesothorium I	6.7 yrs	3.28×10^{-9}	β^-	0.053	0.03		0.0043	4.74×10^{-10}
				γ					
$^{228}_{89}Ac$	Mesothorium II	6.13 hrs	3.141×10^{-5}	α	4.54	0.0581	3.034		
				β	1.0	0.0795		0.40	
				γ	1.6	0.1294		0.73	
					1.85	0.184		0.87	4.95×10^{-14}
					2.16	0.2497		1.04	
						0.3190			
						0.338			
						0.408			
						0.462			
						0.915			
						0.970			
$^{228}_{90}Th$	Radiothorium	1.9 yrs	1.156×10^{-8}	α	0.0658	0.0035			
				γ	0.0689	0.0833			
				e^-	0.0797	0.0868			
					0.0804	0.133			
					0.0833	0.172			
					0.0840	0.216			
					5.173(0.2%)		3.693		1.34×10^{-10}
					5.208(0.4%)		3.731		
					5.338(28%)		3.874		
					5.421(71%)		3.967		

Appendix 1 (*continued*)

Nuclide	Historical name	Period of half-life, $T_{1/2}$	Radioactive constant λ, s^{-1}	Character of radiation	Energy, MeV		Free path of α-particles in the air under normal pressure, cm	Maximum free path of β-particles, g/cm²	Equilibrium mass ratio
					particles	γ-rays			
$^{224}_{88}Ra$	Thorium X	3.64 days	2.204×10^{-6}	α	5.184(0.4%)	0.0301	3.705		
				γ	5.448(4.6%)	0.0428	3.997		6.93×10^{-13}
					5.681(95%)	0.1422	4.267		
						0.2267			
$^{220}_{86}Rn$	Radon	55.0 s	1.2718×10^{-2}	α	6.278		4.982		1.18×10^{-16}
$^{216}_{84}Po$	Thorium A	0.15 s	4.387	α (~100%) β^- (0.014%)	5.996		4.636		3.36×10^{-19}
$^{216}_{85}At$	Astatine	3×10^{-4} s	2.3×10^{2}	α	7.79		7.027		8.96×10^{-25}
						0.1130		0.105	
$^{212}_{82}Pb$	Thorium B	10.64 hrs	1.816×10^{-5}	β^-	0.355	0.1151		0.200	
				γ	0.589	0.1354			7.96×10^{-14}
						0.1641			
						0.1765			
						0.2381			
						0.2505			

$^{212}_{82}Pb$	Thorium B					0.3001			
	(continued)					0.0398			
						0.1243			
				α	5.481(0.016%)	0.1446	4.035		
				(33.7%)	5.603(1.1%)	0.1624	4.175		
					5.662(0.16%)	0.2873	4.245		
					5.765(18%)	0.3271	4.366		
$^{212}_{83}Bi$	Thorium C	1.09 hrs	1.766×10^{-4}	β^-	6.047(69.78%)	0.4316	4.700		8.18×10^{-15}
				(66.3%)					
					6.086(27.25%)	0.4517	4.748		
					0.45(2%)	0.4708		0.135	
					0.62(4%)	0.4908		0.210	
					2.20(~85%)	0.6157		1.07	
					1.55(9%)	0.7195			
						1.623			
						1.797			
$^{212}_{84}Po$	Thorium C′	2.9×10^{-7} s	3.15×10^{6}	α	8.776(~100%)		8.528		
					9.489(0.004%)	9.695			3.04×10^{-25}
					10.417(0.002%)		11.309		
					10.536(0.017%)		11.523		
$^{208}_{81}Tl$	Thorium C″	3.1 min	3.73×10^{-3}	β^-	1.805	0.0405			
				γ		0.2109			
						0.233		0.85	
						0.2518			
						0.2767			
						0.3012			
						0.3227			1.28×10^{-16}
						0.3300			

Appendix 1 (*concluded*)

Nuclide	Historical name	Period of half-life, $T_{1/2}$	Radioactive constant λ, s^{-1}	Character of radiation	Energy, MeV		Free path of α-particles in the air under normal pressure, cm	Maximum free path of β-particles, g/cm^2	Equilibrium mass ratio
					particles	γ-rays			
$^{208}_{81}Tl$	Thorium C′′ (*continued*)					0.5110 0.5832 0.1118 2.166 2.61425 3.324			
$^{208}_{82}Pb$	Thorium D	stable							

According to Baranov B.T., Titayeva N.A. Radiogeology, 1973.

Appendix 2

Auxiliary Data for the Estimation of Decay and Accumulation of Radionuclides

$t/T_{1/2}$	$e^{-\lambda t}$	$1 - e^{-\lambda t}$	$t/T_{1/2}$	$e^{-\lambda t}$	$1 - e^{-\lambda t}$
0.02	0.985	0.015	1.25	0.420	0.580
0.04	0.172	0.028	1.30	0.406	0.594
0.06	0.959	0.041	1.35	0.392	0.608
0.08	0.946	0.054	1.40	0.379	0.621
0.10	0.933	0.067	1.45	0.366	0.634
0.12	0.920	0.080	1.50	0.354	0.646
0.14	0.907	0.093	1.60	0.330	0.670
0.16	0.894	0.106	1.70	0.308	0.692
0.18	0.882	0.118	1.80	0.288	0.712
0.20	0.870	0.130	1.90	0.268	0.732
0.22	0.858	0.142	2.00	0.250	0.750
0.24	0.846	0.154	2.10	0.233	0.767
0.26	0.835	0.165	2.20	0.218	0.782
0.28	0.823	0.177	2.30	0.203	0.797
0.30	0.812	0.188	2.40	0.190	0.810
0.32	0.801	0.199	2.50	0.177	0.823
0.34	0.790	0.210	2.60	0.165	0.835
0.36	0.779	0.221	2.70	0.154	0.846
0.38	0.769	0.231	2.80	0.144	0.856
0.40	0.758	0.242	2.90	0.134	0.866
0.42	0.748	0.252	3.00	0.125	0.875
0.44	0.737	0.263	3.20	0.109	0.891
0.46	0.727	0.273	3.40	0.095	0.905
0.48	0.717	0.283	3.60	0.083	0.917
0.50	0.707	0.293	3.80	0.072	0.928
0.55	0.683	0.317	4.00	0.062	0.938
0.60	0.659	0.341	4.20	0.054	0.946
0.65	0.637	0.363	4.40	0.647	0.953
0.70	0.615	0.385	4.60	0.041	0.959
0.75	0.594	0.406	4.80	0.036	0.964
0.80	0.574	0.426	5.00	0.031	0.969
0.85	0.555	0.445	5.20	0.027	0.973
0.90	0.536	0.464	5.40	0.024	0.976
0.95	0.518	0.482	5.60	0.021	0.979

Appendix 2 (*concluded*)

$t/T_{1/2}$	$e^{-\lambda t}$	$1 - e^{-\lambda t}$	$t/T_{1/2}$	$e^{-\lambda t}$	$1 - e^{-\lambda t}$
1.00	0.500	0.500	5.80	0.018	0.982
1.05	0.483	0.517	6.00	0.016	0.984
1.10	0.467	0.533	7.00	0.008	0.992
1.15	0.451	0.549	8.00	0.004	0.996
1.20	0.435	0.565	9.00	0.002	0.998
			10.00	0.001	0.999

Note: t—time of accumulation or decay of radionuclide;

T—half-life of radionuclide;

λ—decay constant of radionuclide.

Appendix 3

Geological Time Scale (according to W. Horland, A. Cox, P. Llewellyn, C. Picton, A. Smith and R. Walters*)

Table 1

Bln. years	**Chronometric**	**Eon**	**Chronostatigraphic Era**
0—		Phanerozoic	Cenozoic
			Mesozoic
			Paleozoic
			Sinian
	Pt_3	Proterozoic	
1.0—	Pt_2		Riphean
2.0—	Pt_1		Huronian
		Archean	
	Ar_3		Randian
3.0—	Ar_2		Swazian
	Ar_1		Isuan
4.0—		Priscoan	
5.0—			Origin of Earth

Table 2

Era	Subera Period Subperiod		Epoch	Age, mln years	Duration, mln years
Cenozoic Cz	Quaternary or Pleistocene		Holocene	0.01	2
			Pleistocene	2.0	
	Tertiary	Neogene Ng	Pliocene	5.1	22.6
			Miocene	24.6	
		Paleogene ₽	Oligocene	38.0	40.4
			Eocene	54.9	
			Paleocene		
Mesozoic Mz	Cretaceous K		K_2 Cenonian	88.5 97.5	79
			K_1 Neocomian	125 144	
	Jurassic J		J_3 Malm	163	69
			J_2 Dogger	188	
			J_1 Lias	213	
	Triassic Tr		Tr_3	231	35
			Tr_2	243	
			Tr_1 Scythian	248	

Table 2 (*continued*)

Era	Subera Period Subperiod	Epoch	Age, mln. years	Duration, mln. years
Paleozoic Pz	Permian P	P_2	258	38
		P_1	286	
	Carboni-ferrous C	C_2 Pennsylvanian	320	34
		C_1 Mississippian	360	40
	Devonian D	D_3	374	48
		D_2	387	
		D_1	408	
	Silurian S	Přidolian	414	30
		Ludlovian	421	
		Wenlockian	428	
		Llandoverian	438	
	Ordovician O	O_3	458	67
		O_2	478	
		O_1	505	
	Cambrian Є	Merioneth	523	85

Table 2 (*concluded*)

Era	**Subera Period Subperiod**	**Epoch**	**Age, mln. years**	**Duration, mln. years**
Paleozoic Pz		St. David's	540	
		Caerfai	590	
Sinean	Vendian		670	80
	Sturtian	U	800	
Riphean R			1650	
	Huronian	H	2100	
			2400	
	Randian	Ran	2800	
	Swazian	Sw	3750	
	Isuan	I	3900	
	Hadean	Hde		

* Horland, W.E., Cox, A.V., Llewellyn, P.G., Picton, C.A.G., Smith, A.G., Walters, R. (assisted by K.E. Fancett). A Geologic Time Scale. Cambridge University Press, Cambridge-New York-New Rochelle-Melbourne-Sydney, 1982.

Appendix 4

The Slope of an Isochron Corresponding to the Model Age (*t*) for the Single-Stage Lead Model Based on the Equation:

$$m = \frac{1}{137.8}\left[\frac{e^{\lambda_{235}T} - e^{\lambda_{235}t}}{e^{\lambda_{238}T} - e^{\lambda_{238}t}}\right]$$

$$\lambda_{238} = 1.55215 \times 10^{-10} \text{ years}^{-1}$$

$$\lambda_{235} = 9.8485 \times 10^{-10} \text{ years}^{-1}$$

Age of Earth $T = 4.55 \times 10^9$ years

Age (*t*) 10^9 years	**Slope (*m*)**	**Age (*t*) 10^9 years**	**Slope (*m*)**
0.0	0.6180	2.2	0.9335
0.2	0.6360	2.4	0.9816
0.4	0.6555	2.6	1.0347
0.6	0.6766	2.8	1.0936
0.8	0.6996	3.0	1.1590
1.0	0.7247	3.2	1.2317
1.2	0.7520	3.4	1.3128
1.4	0.7819	3.6	1.4034
1.6	0.8146	3.8	1.5047
1.8	0.8505	4.0	1.6184
2.0	0.8900	4.5	1.9681

According to G. Faure, Nuclear Geology, 1986.

Appendix 5

Prefixes Used with Units

Prefix	Symbol	Unit of multiplier
Tera	T	10^{12}
Giga	G	10^{9}
Mega	M	10^{6}
Kilo	k	10^{3}
Hecto	h	10^{2}
Deca	dk	10^{1}
Deci	d	10^{-1}
Centi	c	10^{-2}
Milli	m	10^{-3}
Micro	μ	10^{-6}
Nano	n	10^{-9}
Piko	p	10^{-12}

Examples:

$1\ \mu Ci = 1 \times 10^{-6}\ Ci$
$1\ pCi = 1 \times 10^{-12}\ Ci$
$1\ MeV = 1 \times 10^{6}\ eV$

INDEX